MONOGRAPHS ON
APPLIED PROBABILITY AND STATISTICS

General Editors

M.S. BARTLETT, F.R.S. *and* D.R. COX, F.R.S.

THE ANALYSIS OF CONTINGENCY TABLES

The Analysis of Contingency Tables

B.S. EVERITT

Lecturer, Biometrics Unit,
Institute of Psychiatry, London

LONDON
CHAPMAN AND HALL

A HALSTED PRESS BOOK
JOHN WILEY & SONS INC., NEW YORK

First published 1977
by Chapman and Hall Ltd
11 *New Fetter Lane, London EC4P 4EE*
Reprinted 1979

© 1977 *B.S. Everitt*

Photosetting by Thomson Press (India) Limited, New Delhi
Printed in Great Britain by
J.W. Arrowsmith Ltd., Bristol 3

ISBN 0 412 14970 2

Distributed in the U.S.A. by Halsted Press,
a Division of John Wiley & Sons, Inc., New York

Library of Congress Cataloging in Publication Data

Everitt, B. S.
  The analysis of contingency tables.

  (Monographs on applied probability and statistics)
  1. Contingency tables. I. Title.
QA277.E9 1977   519.5'3   77-4244
ISBN 0-470-99144-5

Contents

Foreword

For several years now my book *Analysing Qualitative Data* has been in need of revision. Since it was first published in 1961, and in part perhaps because of it, a great deal of new and interesting work on the analysis of contingency tables has been published. Mr. Brian Everitt kindly undertook to do the revision but, when he came to review recent literature, it became apparent that a mere renovation of the original text would not be enough; the amount of new work was not only extensive but also made obsolete many of the older methods. In consequence, and with the agreement of the publishers, it was decided that the revised version should in effect be a new book.

That it is so is not strikingly evident in the first two chapters of the present text which, by way of introduction, cover old ground. Thereafter, the increased scope of new methods becomes abundantly apparent. This can be illustrated by a single example. When the literature up to 1961 was reviewed the big disappointment was the paucity and inadequacy of methods then available for the analysis of multidimensional tables, and they are the rule rather than the exception in research work in the social sciences. This serious deficiency has since been met to a truly gratifying extent, not only by the extension of methods for testing more searching hypotheses, but also by the development of methods for fitting log-linear models to multidimensional frequency data and the description of these data in parametric terms. The latter topic is very lucidly described by Everitt in Chapter 5 of this text. Indeed this chapter presents the main challenge to the reader.

Analysing Qualitative Data has been a very popular book amongst research workers for over a decade but the time has come to supplement it by a more modern text. This book by Everitt fills that role and I am confident that it will prove to be as popular as its predecessor.

A.E. Maxwell
Institute of Psychiatry
University of London

January, 1976

Preface

The present text arises from a plan to revise Professor A.E. Maxwell's well known book *Analysing Qualitative Data*. After a review of the current literature it was obvious that major changes would need to be made to that work to bring it up to date, and so, with Professor Maxwell's approval and encouragement, it was decided that I should attempt to produce almost a new book, with the present title. Readers familiar with the original text will see that the first three chapters of *The Analysis of Contingency Tables* are very similar in content to parts of the first six chapters of that work. However, the last three chapters of the current work deal with topics not covered in the original, namely the analysis of multidimensional contingency tables.

It was intended that the present text should be suitable for a similar group of people as the original, namely research workers in psychiatry, the social sciences, psychology, etc., and for students undergoing courses in medical and applied statistics. This has meant that the mathematical level of the book has been kept deliberately low.

While in preparation the manuscript was read by Professor Maxwell whose help and encouragement were greatly appreciated. Thanks are also due to the Biometrika trustees for permission to reproduce the table of chi-squared values given in Appendix A, and to Mrs. B. Lakey for typing the manuscript.

B.S. Everitt
Institute of Psychiatry
University of London

May, 1976

CHAPTER ONE

Contingency tables and the chi-square test

1.1. Introduction

This book is primarily concerned with methods of analysis for *frequency data* occurring in the form of cross-classification or *contingency tables*. In this chapter we shall commence by defining various terms, introducing some nomenclature, and describing how such data may arise. Later sections describe the chi-square distribution, and give a numerical example of testing for *independence* in a contingency table by means of the *chi-square test*.

1.2. Classification

It is possible to classify the members of a *population* – a generic terms denoting any well defined class of people or things – in many different ways. People, for instance, may be classified into male and female, married and single, those who are eligible to vote and those who are not, and so on. These are examples of dichotomous classifications. Multiple classifications also are common, as when people are classified into left-handed, ambidextrous, and right-handed, or, for the purpose of say a gallup poll, into those who intend to vote (a) Conservative, (b) Labour, (c) Liberal, (d) those who have not yet made up their minds, and (e) others. We shall be primarily interested in classifications whose categories are *exhaustive* and *mutually exclusive*. A classification is exhaustive when it provides sufficient categories to accommodate all members of the population. The categories are mutually exclusive when they are so defined that each member of the population can be correctly allocated to one, and only one, category. At first sight it might appear that the requirement that a classification be exhaustive is very restrictive. We might, for example, be interested in carrying out a gallup poll, not on the voting intentions of the electorate as a whole, but only on those of university students. The difficulty is resolved if the definition of a population is recalled. The statistical definition of the word is more fluid than its definition in ordinary usage, so it is quite in order to define the population in question as 'all university students eligible to vote'. Categories too are adjustable and may often be altered or combined; for instance, in the voting example it is unlikely that much information would be lost by amalgamating categories (d) and (e).

When the population is classified into several categories we may then 'count' the number of individuals in each category. These 'counts' or frequencies are the type of data with which this book will be primarily concerned. That is, we shall be dealing with *qualitative* data rather than with *quantitative* data obtained from measurement of continuous variables such as height, temperature, and so on.

In general, of course, information from the whole population is not available and we must deal with only a *sample* from the population. Indeed, one main function of statistical science is to demonstrate how valid inferences about some population may be made from an examination of the information supplied by a sample. An essential step in this process is to ensure that the sample taken is a representative (unbiased) one. This can be achieved by drawing what is called a *random sample*, that is one in which each member of the population in question has an equal chance of being included. The concept of random sampling is discussed in more detail in Chapter 2.

1.3. Contingency tables

The main concern of this book will be the analysis of data which arise when a sample from some population is classified with respect to two or more qualitative variables. For example, Table 1.1 shows a sample of 5375 tuberculosis deaths classified with respect to two qualitative variables, namely sex and type of tuberculosis causing death. (Note that the categories of these variables as given in the table are both exhaustive and mutually exclusive.)

A table such as Table 1.1 is known as a *contingency table*, and this 2×2 example (the members of the sample having been dichotomized in two different ways) is its simplest form. Had the two variables possessed multiple rather than dichotomous categories the table would have had more cells than the four shown. The entries in the

TABLE 1.1. Deaths from tuberculosis.

	Males	Females	Total
Tuberculosis of respiratory system	3534	1319	4853
Other forms of tuberculosis	270	252	522
Tuberculosis (all forms)	3804	1571	5375

cells for these data are frequencies. These may be transformed into proportions or percentages but it is important to note that, in whatever form they are presented, the data were originally frequencies or counts rather than continuous measurements. Of course, continuous data can often be put into discrete form by the use of intervals on a continuous scale. Age, for instance, is a continuous variable, but if people are classified into different age groups the intervals corresponding to these groups can be treated as if they were discrete units.

Since Table 1.1 involves only two variables it may be referred to as a *two-dimensional* contingency table; in later chapters we shall be concerned with three-dimensional and higher contingency tables which arise when a sample is classified with respect to *more* than two qualitative variables.

1.4. Nomenclature

At this point we shall introduce some general notation for two-dimensional tables. Later, when dealing with higher dimensional tables, this notation will be extended.

The general form of a two-dimensional contingency table is given in Table 1.2, in which a sample of N observations is classified with respect to two qualitative variables, one having r categories and the other having c categories. It is known as an $r \times c$ *contingency table*.

The observed frequency or count in the ith category of the row variable and the jth category of the column variable, that is the frequency in the ijth cell of the table, is represented by n_{ij}. The total number of observations in the ith category of the row variable is denoted by $n_{i.}$ and the total number of observations in the jth category of the column variable by $n_{.j}$. These are known as *marginal totals*, and in terms of the cell frequencies, n_{ij}, are given by:

TABLE 1.2. General form of a two-dimensional contingency table.

		Columns (Variable 2)							
		1	2	·	·	·	·	c	Total
Rows (Variable 1)	1	n_{11}	n_{12}	·	·	·	·	n_{1c}	$n_{1.}$
	2	n_{21}							$n_{2.}$
	r	n_{r1}						n_{rc}	$n_{r.}$
Total		$n_{.1}$	$n_{.2}$					$n_{.c}$	$n_{..} = N$

$$n_{i.} = n_{i1} + n_{i2} + \ldots + n_{ic}$$

$$= \sum_{j=1}^{c} n_{ij} \tag{1.1}$$

$$n_{.j} = n_{1j} + n_{2j} + \ldots + n_{rj}$$

$$= \sum_{i=1}^{r} n_{ij} \tag{1.2}$$

Similarly

$$n_{..} = \sum_{i=1}^{r} \sum_{j=1}^{c} n_{ij} \tag{1.3}$$

$$= \sum_{i=1}^{r} n_{i.} = \sum_{j=1}^{c} n_{.j} \tag{1.4}$$

$n_{..}$ represents the total number of observations in the sample and is usually denoted by N.

This notation is generally known as *dot notation*, the dots indicating summation over particular subscripts.

In the case of the data shown in Table 1.1 we have:

(I) $r = c = 2$, that is both variables have two categories;

(II) $n_{11} = 3534$, $n_{12} = 1319$, $n_{21} = 270$, and $n_{22} = 252$ are the cell frequencies;

(III) $n_{1.} = 4853$ and $n_{2.} = 522$ are the *row* marginal totals, that is the total number of deaths from the two types of tuberculosis;

(IV) $n_{.1} = 3804$ and $n_{.2} = 1571$ are the *column* marginal totals, that is the total number of males and females in the sample;

(V) $N = 5375$ is the total number of observations in the sample.

1.5. Independent classifications–association

Having examined the type of data with which we are concerned, we now need to consider the questions of interest about such data. In general the most important question is whether the qualitative variables forming the contingency table are *independent* or not. To answer this question, it is necessary to get clear just what independence between the classifications would entail. In the case of a 2×2 table this is relatively easy to see. For example, returning to the data of Table 1.1, it is clear that, if the form of tuberculosis from which people die is independent of their sex, we would expect the proportion of males that died from tuberculosis of the respiratory system to be equal to the proportion of females that died from the same cause. If these proportions differ, death from tuberculosis of the respiratory system tends to be *associated* more with one of the sexes than with the other. (Of course, the two proportions might be expected to differ in some measure due solely to chance factors of sampling, and for other reasons which might be attributed to random causes; what we shall need to ascertain is whether or not the observed difference between the proportions is *too* large to be attributed to such causes, and for this we will require the test that is discussed in the following section.)

Having seen, intuitively, that independence in a 2×2 table implies the equality of two proportions, let us now examine slightly more formally what the concept implies for the general $r \times c$ contingency table. First suppose that, in the population from which the sample is to be taken, the probability of an observation belonging to the ith category of the row variable *and* the jth category of the column variable is represented by p_{ij}; consequently the frequency, F_{ij}, to be expected in the ijth cell of the table resulting from sampling N individuals, is given by:

$$F_{ij} = Np_{ij} \tag{1.5}$$

[Readers familiar with mathematical expectation and probability distributions will recognize that $F_{ij} = E(n_{ij})$, under the assumption that the observed frequencies follow a *multinomial distribution* with probability values p_{ij}; see, for example, Mood and Graybill, 1963, Ch. 3.]

Now, let $p_{i.}$ represent the probability, in the population, of an observation belonging to the ith category of the row variable (in this case with no reference to the column variable), and let $p_{.j}$ represent the corresponding probability for the jth category of the column variable. Then, from the multiplication law of probability, independence between the two variables, *in the population*, implies that:

$$p_{ij} = p_{i.} p_{.j} \qquad (1.6)$$

In terms of the frequencies to be expected in the contingency table, independence is therefore seen to imply that:

$$F_{ij} = N p_{i.} p_{.j} \qquad (1.7)$$

However, the reader might ask in what way this helps since the independence of the two variables has only been defined in terms of *unknown* population probability values. The answer is that these probabilities may in fact be *estimated* very simply from the observed frequencies, and it is easy to show that the 'best' estimates $\hat{p}_{i.}$ and $\hat{p}_{.j}$ of the probabilities $p_{i.}$ and $p_{.j}$ are based upon the relevant marginal totals of observed values; that is:

$$\hat{p}_{i.} = \frac{n_{i.}}{N} \text{ and } \hat{p}_{.j} = \frac{n_{.j}}{N} \qquad (1.8)$$

(These are *maximum likelihood estimates*; see Mood and Graybill, Ch. 12.) The use of the estimates of $p_{i.}$ and $p_{.j}$ given in equation (1.8) allows us now to estimate the frequency to be expected in the *ij*-cell of the table if the two variables were independent. Inspection of equation (1.7) shows that this estimate, which we shall represent as E_{ij}, is given by:

$$E_{ij} = N \hat{p}_{i.} \hat{p}_{.j}$$

$$= N \frac{n_{i.} n_{.j}}{N \, N} = \frac{n_{i.} n_{.j}}{N} \qquad (1.9)$$

When the two variables *are* independent the frequencies estimated using formula (1.9) and the observed frequencies should differ by amounts attributable to chance factors only. If however the two variables are *not* independent we would expect larger differences to arise. Consequently it would seem sensible to base any test of the independence of the two variables forming a two-dimensional contingency table on the size of the differences between the two sets of frequencies, n_{ij} and E_{ij}. Such a test is discussed in the following section. (In latter parts of this text the estimated expected frequencies, E_{ij}, will, when there is no danger of confusing them with the frequencies, F_{ij}, often be referred to simply as 'expected values'.)

1.6. Chi-square test

In the preceding section the concept of the independence of two variables was discussed. To test for independence it was indicated that we need to investigate the truth of the *hypothesis*:

$$p_{ij} = p_{i.} p_{.j} \qquad (1.10)$$

In general this hypothesis will be referred to as the *null hypothesis* and denoted by the symbol H_0.

It was also pointed out that the test should be based upon the differences between the estimated values of the frequencies to be expected when H_0 is true (that is the E_{ij}) and the observed frequencies (that is the n_{ij}). Such a test, first suggested by Pearson (1904), uses the statistic χ^2 given by:

$$\chi^2 = \sum_{i=1}^{r} \sum_{j=1}^{c} \frac{(n_{ij} - E_{ij})^2}{E_{ij}} \qquad (1.11)$$

It is seen that the magnitude of this statistic depends on the values of the differences $(n_{ij} - E_{ij})$. If the two variables are independent these differences will be less than otherwise would be the case; consequently χ^2 will be smaller when H_0 is true than when it is false. Hence what is needed is a method for deciding on values of χ^2 which should lead to acceptance of H_0 and those which should lead to its rejection. Such a method is based upon deriving a *probability distribution* for χ^2 under the assumption that the hypothesis of independence is true. Acceptance or rejection of the hypothesis is then based upon the probability of the obtained χ^2 value; values with 'low' probability lead to rejection of the hypothesis, others to its acceptance. This is the normal procedure for deriving *significance tests* in statistics. In general a 'low' probability is taken to be a value of 0.05 or 0.01, and is referred to as the *significance level* of the test. (For a more detailed discussion of significance level and related topics, see Mood and Graybill, Ch. 12.)

By assuming that the observed frequencies have a particular distribution, namely a multinomial distribution, and by further assuming that the expected frequencies are not too small (see page 45), the statistic χ^2 may be shown to have approximately a *chi-square distribution*. The test of the hypothesis of independence may now be performed by comparing the obtained value of χ^2 with the tabulated values of this distribution.

1.7. Chi-square distribution

We shall assume that readers are familiar with the *normal distribution*, accounts of which are given in most statistical text books (see again Mood and Graybill, Ch. 6). The chi-square distribution arises from it as the probability distribution of the sums of squares of a number of independent variables, z_i, each of which has a standard normal distribution, that is one with mean zero and standard deviation unity. The form of the distribution depends upon the number of independent variates involved. For example, a chi-square variable (χ^2) formed by the sum of squares of v independent standard normal variables, namely:

$$\chi^2 = z_1{}^2 + z_2{}^2 + \ldots + z_v{}^2 \tag{1.12}$$

has a distribution depending only on v. Diagrams showing the different shapes the distribution takes, for varying values of v, are given in many text-books (for example, Hays, 1973, Ch. 11). In general the number of independent variates forming the chi-square variable is known as the *degrees of freedom*; in the above case we would speak of a chi-square with v degrees of freedom (d.f.).

The mathematical expression for the chi-square distribution need not be discussed here since tabulated values (Appendix A) are available. They give all the information necessary for deciding whether the value of χ^2 obtained for some contingency table should lead us to accept or reject the hypothesis of independence. Firstly, however, we need to know the degrees of freedom of χ^2; this depends on the number of categories of each variable forming the table, and its value is derived in the following section. Knowing v, we examine the tables of the chi-square distribution with v degrees of freedom, for some *a priori* determined significance level, say α (usually 0.05 or 0.01), and find the requisite value of chi-square. If χ^2 is greater than the tabulated value, denoted by $\chi_v{}^2$, at the α level, it indicates that the result obtained would be expected to occur by chance very rarely (less than $100\,\alpha\%$ of the time); consequently it is indicative of a real departure from independence, and hence we are led to reject our null hypothesis, H_0.

1.8. Degrees of freedom for a two-dimensional contingency table

The statistic for testing the independence of the two variables forming a contingency table has already been seen to be:

$$\chi^2 = \sum_{i=1}^{r} \sum_{j=1}^{c} \frac{(n_{ij} - E_{ij})^2}{E_{ij}} \tag{1.13}$$

The degrees of freedom of the chi-square distribution which approximates the distribution of χ^2 when the hypothesis of independence is true, is simply the number of independent terms in (1.13), *given that the row and column marginal totals of the table are fixed.* The *total* number of terms in (1.13) is $r \times c$, that is the number of cells in the table. Some of these terms are however determined by knowledge of the row and column totals. For example, knowing the r row totals fixes r of the frequencies n_{ij}, one in each row, and hence determines r of the terms in (1.13). The number of independent terms in (1.13) is thus reduced to $(rc - r)$. If we assume that the frequency fixed by each row total is that in the last column, we see that, of the c column totals, only the first $(c - 1)$ remain to be considered. Each of these fixes one frequency in the body of the table and consequently reduces the number of independent terms by one. Hence we are left with $rc - r - (c - 1)$ independent terms in (1.13). This gives the degrees of freedom of χ^2:

$$\text{d.f.} = rc - r - (c - 1)$$
$$= (r - 1)(c - 1) \tag{1.14}$$

(The degrees of freedom of a contingency table may also be regarded as the number of cells of the table which may be filled arbitrarily when the marginal totals are fixed.)

1.9. Numerical example

To aid the reader in understanding the discussion so far, let us examine again the data shown in Table 1.1. The hypothesis we wish to test is that the form of tuberculosis from which people die is independent of sex – which is another way of saying that the two classifications are independent.

The first step in the calculation of χ^2 is to compute the expected values using formula (1.9). For example, E_{11} is given by:

$$E_{11} = \frac{4853 \times 3804}{5375} \tag{1.15}$$

$$= 3434.6$$

We may now arrange the rest of the calculations as in Table 1.3.

TABLE 1.3. Calculating χ^2 for the data of Table 1.1

(1)	(2)	(3)	(4)	(5)
n_{ij}	E_{ij}	$(n_{ij} - E_{ij})$	$(n_{ij} - E_{ij})^2$	$(n_{ij} - E_{ij})^2/E_{ij}$
3534	3434.6	99.4	9880.36	2.88
1319	1418.4	− 99.4	9880.36	6.97
270	369.4	− 99.4	9880.36	26.75
252	152.6	99.4	9880.36	64.75
5375	5375.0	0.0		$\chi^2 = 101.35$

The differences $(n_{ij} - E_{ij})$ are obtained, a check being that they add to zero. Next the differences are squared and each squared difference is divided by the expected frequency in its own row. These values appear in column (5) of the table and their sum gives the required value of χ^2.

To assess whether a value of $\chi^2 = 101.35$ leads us to accept or reject our hypothesis of independence, we refer to the appropriate chi-square distribution, that is the distribution with the degrees of freedom of Table 1.1. For our data $r = c = 2$, and therefore using formula (1.14) gives unity as the number of degrees of freedom. This shows that only one cell in the table may be fixed arbitrarily, given the row and column totals. If we perform our test at the 0.05 or 5% significance level, the requisite value of chi-square with 1 d.f., obtained from Appendix A, is 3.84. Since our value of χ^2 is far greater than this we conclude that the variables are not independent but are 'associated'. Since we are dealing with a 2×2 table this may be interpreted as meaning that the proportion of males who died from tuberculosis of the respiratory system, namely $3534/3804 = 0.929$, is significantly different from the proportion of females, namely $1319/1571 = 0.840$, who died from the same cause.

It is well to note that the finding of a significant association by means of the chi-square test does not necessarily imply any causal relationship between the variables involved, although it does indicate that the reason for this association is worth investigating.

1.10. Summary

In this chapter the type of data with which this book is primarily concerned has been described. Testing for the independence of two qualitative variables by means of a chi-square test was introduced.

The chi-square test is central to this text, and in subsequent chapters many examples of its use in investigations in psychology, psychiatry, and social medicine will be given. These sciences are as yet in the early stages of development, and studies in them are still characterized by a search for the variables basic to them. This search is often one for general relationships and associations— however amorphous they may appear at first—between the phenomena being studied, and here the chi-square test is often helpful.

2×2 Contingency tables

2.1. Introduction

The simplest form of contingency table, namely that arising from two dichotomous variables and known as a 2×2 table, has been introduced in the preceding chapter. In this chapter we shall deal with such tables in greater detail.

Data which form 2×2 tables occur very frequently in social science and educational and psychiatric research. Such data may arise in several ways. For instance, they can occur when a total of say N subjects is sampled from some population and each individual is classified according to two dichotomous variables. For example, in an investigation of the relationship between age and smoking in males, one hundred males might be sampled, and dichotomized with respect to age, say above or below forty, and with respect to amount of smoking, say above and below twenty cigarettes a day. A (hypothetical) set of data is shown in Table 2.1.

Again such data may arise in an investigation where we sample a *predetermined* number of individuals in each of the categories of one of the variables, and for each sample assess the number of individuals in each of the categories of the second variable. For example, in an investigation into the frequency of side-effects, say

TABLE 2.1. Smoking and age example.

		Age		
		Under 40	Over 40	
Amount of smoking	Less than 20/day	50	15	65
	More than 20/day	10	25	35
		60	40	100

TABLE 2.2. Side-effects and drug example.

		Side-effect (Nausea) Present	Absent	
Treatment	Drug given	15	35	50
	Placebo given	4	46	50
		19	81	100

nausea, with a particular drug, fifty subjects may be given the drug, fifty subjects given a placebo, and the number of subjects suffering from nausea assessed in each sample. Table 2.2 shows a possible outcome.

The analysis of such tables is by means of the chi-square test described in Chapter 1, a simplified form of which is available for 2 × 2 tables as indicated in the following section.

2.2. Chi-square test for a 2 × 2 table

The general 2 × 2 table may be written in the form shown in Table 2.3.

The usual expression for computing the statistic χ^2, that is:

$$\sum \frac{(\text{Observed frequency} - \text{Expected frequency})^2}{\text{Expected frequency}} \tag{2.1}$$

reduces, for the general 2 × 2 table, to the following simplified form:

$$\chi^2 = \frac{N(ad - bc)^2}{(a+b)(c+d)(a+c)(b+d)} \tag{2.2}$$

TABLE 2.3. General 2 × 2 contingency table.

		Variable A Category 1	Category 2	
Variable B	Category 1	a	b	$a+b$
	Category 2	c	d	$c+d$
		$a+c$	$b+d$	$N = a+b+c+d$

As was seen in Chapter 1, the significance of this statistic is judged by referring it to the tabulated χ^2 values with one degree of freedom. For example, applying formula (2.2) to the data of Table 2.2 gives:

$$\chi^2 = \frac{100 \times (15 \times 46 - 35 \times 4)^2}{50 \times 50 \times 19 \times 81} = 7.86$$

At the 5% level the tabulated χ^2 value for 1 d.f. is 3.84. Our computed value of χ^2 is greater than this value and we are led to suspect the truth of our hypothesis that the occurrence of side-effects is independent of the treatments (drug and placebo) involved. The result may be interpreted as indicating that the proportion of people who suffer from nausea when treated with the drug is different from the proportion of people who suffer from nausea when given a placebo. These proportions, estimated from Table 2.2, are 0.30 and 0.08.

2.3. Yates's continuity correction

In deriving the distribution of the statistic χ^2 essentially we are employing a *continuous* probability distribution, namely the chi-square distribution, as an approximation to the *discrete* probability distribution of observed frequencies, namely the multinomial distribution. To improve the approximation, Yates (1934) suggested a correction which involves subtracting 0.5 from the positive discrepancies (Observed − Expected), and adding 0.5 to the negative discrepancies, before these values are squared. This correction may be incorporated directly into formula (2.2), which then becomes:

$$\chi^2 = \frac{N(|ad - bc| - 0.5N)^2}{(a + b)(c + d)(a + c)(b + d)} \tag{2.3}$$

It is known as a chi-square value corrected for continuity. In formula (2.3) the term $|ad - bc|$ means 'the absolute value of $(ad - bc)$', that is the numerical value of the expression irrespective of its sign.

Recently there has been some discussion of the merits of applying Yates's correction. Conover (1968, 1974) questions its routine use in all cases, but Mantel and Greenhouse (1968), Fleiss (1973), and Mantel (1974) reject Conover's arguments. In general the evidence for applying the correction seems convincing, and hence its use is recommended. If the sample size is reasonably large the correction, of course, will have little effect on the value of χ^2.

2.4. Small expected frequencies – Fisher's exact test for 2 × 2 tables

As mentioned briefly in Chapter 1, one of the assumptions made when deriving the chi-square distribution, as an approximation to the distribution of the statistic χ^2, is that the expected frequencies should not be too small since otherwise the approximation may not be good. We return to the problem in the following chapter, but in the case of 2 × 2 tables with 'small' expected frequencies, say 5 or less, we may employ *Fisher's exact test* as an alternative to the chi-square test.

2.4.1. *Fisher's test for* 2 × 2 *tables*

Fisher's exact test for a 2 × 2 contingency table does not use the chi-square approximation at all. Instead the exact probability distribution of the observed frequencies is used. For fixed marginal totals the required distribution is easily shown to be that associated with sampling without replacement from a finite population, namely a *hypergeometric distribution* (see Mood and Graybill, Ch. 3). Assuming that the two variables are independent, the probability (P) of obtaining any particular arrangement of the frequencies *a*, *b*, *c*, and *d* (Table 2.3), when the marginal totals are as given, is:

$$P = \frac{(a+b)!\,(c+d)!\,(a+c)!\,(b+d)!}{a!\,b!\,c!\,d!\,N!} \qquad (2.4)$$

where *a*! – read '*a* factorial' – is the shorthand method of writing the product of *a* and all the whole numbers less than it, down to unity; for example:

$$5! = 5 \times 4 \times 3 \times 2 \times 1 = 120$$

(By definition the value of 0! is unity.) Fisher's test now employs formula (2.4) to find the probability of the arrangement of frequencies actually obtained, *and* that of every other arrangement giving as much or more evidence for association, always keeping in mind that the marginal totals are to be regarded as fixed. The sum of these probabilities is then compared with the chosen significance level α; if it is greater than α we have no evidence of any association between the variables; if it is less than α we conclude that the hypothesis of independence should be rejected and therefore that there is a significant association between them. A numerical example will help to clarify the procedure.

TABLE 2.4. The incidence of 'suicidal feelings' in samples of psychotic and neurotic patients.

| | Type of patient | | |
	Psychotics	Neurotics	
Suicidal feelings	2 (4)	6 (4)	8
No suicidal feelings	18 (16)	14 (16)	32
	20	20	40

2.4.2. Numerical example of Fisher's exact test for 2 × 2 tables

In a broad general sense psychiatric patients can be classified as psychotics or neurotics. A psychiatrist, whilst studying the symptoms of a random sample of twenty from each of these populations, found that, whereas six patients in the neurotic group had suicidal feelings, only two in the psychotic group suffered in this way, and he wished to test if there is an association between the two psychiatric groups and the presence or absence of suicidal feelings. The data are shown in Table 2.4. Our hypothesis in this case is that the presence or absence of suicidal feelings is independent of the type of patient involved, or, equivalently, that the proportion of psychotics with suicidal feelings is equal to the proportion of neurotics with this symptom.

The expected frequencies on the hypothesis of independence are shown in parentheses in Table 2.4; we see that two of them are below 5 and consequently we shall test our hypothesis by means of Fisher's exact test rather than by the chi-square statistic. Using formula (2.4) we first find the probability of the observed table; it is:

$$P_2 = \frac{8! \times 32! \times 20! \times 20!}{2! \times 6! \times 18! \times 14! \times 40!} = 0.095760$$

(The subscript to P refers to the smallest of the frequencies a, b, c, d; in this case it is 2.)

Returning to Table 2.4, and keeping in mind that the marginal frequencies are to be taken as fixed, the frequencies in the body of the table can be arranged in two ways both of which would represent, had they been observed, more extreme discrepancies between the groups with respect to the symptom. These arrangements are shown in Table 2.5.

TABLE 2.5. More extreme cell frequencies than those observed.

(a)			(b)		
1	7	8	0	8	8
19	13	32	20	12	32
20	20	40	20	20	40

Substituting in turn the values in Table 2.5(a) and 2.5(b) in formula (2.4) we obtain:

$$P_1 = 0.020160, \text{ for Table 2.5(a)}$$

$$P_0 = 0.001638, \text{ for Table 2.5(b)}$$

Therefore the probability of obtaining the observed result (that is Table 2.4), or one more suggestive of a departure from independence, is given by:

$$P = P_2 + P_1 + P_0$$

$$= 0.095760 + 0.020160 + 0.001638$$

$$= 0.117558$$

This is the probability of observing, amongst the eight patients suffering from suicidal feelings, that two or fewer are psychotics, when the hypothesis of the equality of the proportions of psychotics and neurotics having the symptom, in the populations from which the samples were taken, is true; it shows that a discrepancy between the groups as large as that obtained might be expected to occur by chance about one in ten times even with methods of classification that were independent. Since its value is larger than the commonly used significance levels (0.05 or 0.01), the data give no evidence that psychotics and neurotics differ with respect to the symptom. Indeed, in this case, since P_2 is itself greater than 0.05, the computation could have ended before evaluating P_1 and P_0.

A significant result from Fisher's test indicates departure from the null hypothesis in a *specific direction*, in contrast to the chi-square test which tests departure from the hypothesis in either direction. In the psychiatric groups example, the former is used to decide whether the proportions of patients in the two groups having suicidal feelings are equal or whether the proportion of psychotics with the symptom is *less* than the proportion of neurotics. The normal chi-square statistic, however, tests whether these proportions are equal or unequal without regard to the direction of the inequality. In other words, Fisher's test is *one-tailed* whereas the chi-square test is *two-tailed*. (For a detailed discussion of one-tailed and two-tailed tests, see Mood and Graybill, Ch. 12.) In the case where the sample sizes in each group are the same (as they are in the above example) the probability obtained from Fisher's test may be doubled to give the equivalent of a two-tailed test. This gives $P = 0.23512$. It is of interest to compare this value with the probability that would be obtained using the chi-square test. First we compute the chi-square statistic for Table 2.4, using formula (2.2) to give:

$$\chi^2 = \frac{40(2 \times 14 - 18 \times 16)^2}{20 \times 20 \times 8 \times 32} = 2.50$$

Using the tables given by Kendall (1952, Appendix, Table 6), the exact value of the probability of obtaining a value of chi-square with 1 d.f. as large as or larger than 2.50 may be found; it is 0.11385. Now, we calculate the chi-square statistic with Yates's continuity correction applied, using formula (2.3):

$$\chi^2 = \frac{40(|28 - 108| - 20)^2}{20 \times 20 \times 8 \times 32} = 1.41$$

In this case the corresponding probability is 0.23572. Since the comparable probability obtained from Fisher's test is 0.23512, the efficacy of Yates's correction is clearly demonstrated.

2.4.3. *Calculating the probabilities for Fisher's exact test*

There are various ways in which the calculations involved in applying Fisher's test may be made easier. Perhaps the most convenient method is by use of the tables given in *Biometrika Tables for Statisticians* (1967), and in Finney *et al.* (1963), which enable the 2×2 table to be examined for significance directly. These tables may be used for values of N up to the order of 50.

If evaluation of the probabilities *is* necessary, the values of factorials are given in Barlow's (1952) and in Fisher and Yates's tables (1957). More convenient to use in most cases are tables of the logarithms of factorials (see, for example, Lindley and Miller, 1953). If these tables are not available, a short-cut method of evaluating the probabilities is described by Feldman and Klinger (1963), which involves the application of the following recursive formula:

$$P_{i-1} = \frac{a_i d_i}{b_{i-1} c_{i-1}} P_i \qquad (2.5)$$

In formula (2.5), P_i is the probability of the observed table of frequencies, the subscript i referring to the smallest frequency in the table. Similarly P_{i-1}, P_{i-2}, etc. are the probabilities of the arrangements obtained when this frequency is reduced to give tables more suggestive of a departure from independence. The terms a_i, d_i, b_{i-1}, and c_{i-1} refer to frequencies in the ith and $(i-1)$th tables. To clarify the use of this technique we shall employ it to re-compute the probabilities required for Table 2.4. First we need to calculate P_2, the probability of the observed table. As before, we do this by using formula (2.4) to give $P_2 = 0.09576$. Now using formula (2.5) the other probabilities, namely P_1 and P_0, may be obtained very simply as follows:

$$P_1 = \frac{2 \times 14}{7 \times 19} P_2 = 0.020160 \text{ (as before)}$$

[2 and 14 are the 'a' and 'd' frequencies in Table 2.4; 7 and 19 are the 'c' and 'b' frequencies in Table 2.5(a).]

$$P_0 = \frac{1 \times 13}{20 \times 8} P_1 = 0.001638 \text{ (as before)}$$

2.4.4. *The power of Fisher's exact test for 2 × 2 tables*

The power of a statistical test (see, for example, Mood and Graybill, Ch. 12) is equal to the probability of rejecting the null hypothesis when it is untrue, or in other words the probability of making a correct decision when applying the test. Obviously we would like the power of any test we use to be as high as possible. Several workers have investigated the power of Fisher's test and have shown that large sample sizes are needed to detect even moderately large differences between the two proportions. For example, Bennett and Hsu (1960), for the data of Table 2.4, where sample sizes of 20 from each group (that is 20 neurotics and 20 psychotics) are involved, show that if we are testing at the 5% level, and the population values of the proportions of people with suicidal feelings in each group are 0.5 and 0.2, then the power takes a value of 0.53. Therefore in almost half the cases of performing the test with these sample sizes we shall conclude that there is *no* difference between the incidence of the symptom amongst psychotic and neurotic patients. The results of Gail and Gart (1973) show further that, in this particular example, a sample of 42 patients from each group would be necessary to attain a power of 0.9 of detecting a difference in the population proportions. To detect small differences between the proportions, the latter authors show that relatively large sample sizes may be needed. For example, if the values of the proportions in the population were 0.8 and 0.6, a sample of 88 individuals from each group would be needed to achieve a power of 0.9, that is to have a 90% chance of detecting the difference.

2.5. McNemar's test for correlated proportions in a 2 × 2 table

One-to-one matching is frequently used by research workers to increase the precision of a comparison. The matching is usually done on variables such as age, sex, weight, I.Q., and the like, information about which can be obtained relatively easily. Two samples matched in a one-to-one way must be thought of as correlated rather than independent; consequently the usual chi-square test is not strictly applicable for assessing the difference between frequencies obtained with reference to such samples.

The appropriate test for comparing frequencies in matched samples is one due to McNemar (1955). As an introduction to it, let us look at Table 2.6 in which the presence or absence of some characteristic or attribute A for two matched samples I and II is shown. As we are concerned with the differences between the two samples, the entries in the N-E and S-W cells of the table do not interest us, since the frequency b refers to matched pairs both of which possess the attribute, while the frequency c refers to pairs both of which do not possess the attribute. The comparison is thus confined to the

TABLE 2.6. Frequencies in matched samples.

		Sample I	
		A absent	A present
Sample II	A present	a	b
	A absent	c	d

frequencies a and d, the former representing the number of matched pairs that possess the attribute if they come from sample I and do not possess it if they come from sample II, while the latter represents pairs for which the converse is the case. Under the hypothesis that the two samples do not differ as regards the attribute, we would expect a and d to be equal, or, to put it another way, the expected values for the two cells each to be $(a + d)/2$. Now if the observed frequencies a and d and their expected frequencies $(a + d)/2$ are substituted in the usual formula for χ^2, that is formula (2.1), we obtain:

$$\chi^2 = \frac{(a - d)^2}{a + d} \tag{2.6}$$

If a correction for continuity is applied, this expression becomes:

$$\chi^2 = \frac{(|a - d| - 1)^2}{a + d} \tag{2.7}$$

This is McNemar's formula for testing for an association in a 2×2 table when the samples are matched; under the hypothesis of no difference between the matched samples with respect to the attribute A, χ^2 has a chi-square distribution with 1 d.f. To illustrate McNemar's test let us now consider an example.

2.5.1. Numerical example of McNemar's test

A psychiatrist wished to assess the effect of the symptom 'depersonalization' on the prognosis of depressed patients. For this purpose 23 endogenous depressed patients, who were diagnosed as being 'depersonalized', were matched one-to-one for age, sex,

TABLE 2.7. Recovery of 23 pairs of depressed patients.

| | | Depersonalized patients | | |
		Not recovered	Recovered	
Patients not depersonalized	Recovered	5	14	19
	Not recovered	2	2	4
		7	16	23

duration of illness, and certain personality variables, with 23 endogenous depressed patients who were diagnosed as not being 'depersonalized'. The numbers of pairs of patients from the two samples who, on discharge after a course of E.C.T., were diagnosed as 'recovered' or 'not recovered' are given in Table 2.7. From this table we see that a is 5 and d is 2. Substituting these values in formula (2.7) gives:

$$\chi^2 = \frac{(|5 - 2| - 1)^2}{(5 + 2)}$$

$$= 0.57$$

With one degree of freedom the value does not reach an acceptable level of significance, so we conclude that 'depersonalization' is not associated with prognosis where endogenous depressed patients are concerned.

2.6. Gart's test for order effects

Let us now consider a further commonly occurring situation in which McNemar's test would be relevant, namely that where two drugs A and B are given to patients on two different occasions and some response of interest is noted. Again we are dealing with correlated rather than independent observations (since the same subject receives both A and B) but in this case a complicating factor would be the order of the drug administration since this might have an appreciable effect on a subject's response. The McNemar test may ignore pertinent and important information on order within pairs. For example, suppose the two drugs are used in the treatment of depression and that they are to be investigated

TABLE 2.8. Number of subjects showing nausea with drugs A and B.

| | | Drug A | | |
		No nausea	Nausea	
Drug B	Nausea	3	9	12
	No nausea	75	13	88
		78	22	100

for possible side-effects, say nausea. The drugs are given to 100 patients and the response 'nausea' or 'no nausea' is recorded, giving the results shown in Table 2.8. Therefore, of these subjects, 75 never had nausea, 13 subjects had nausea with A but not with B, 3 had nausea with B but not with A, and 9 had nausea with both drugs. A McNemar test of these data gives, using formula (2.7):

$$\chi^2 = \frac{(|3 - 13| - 1)^2}{16}$$

$$= 5.06$$

Referring this to a chi-square with 1 d.f. at the 5% level, namely 3.84, we see that our value of χ^2 is significant, and we would conclude that the incidence of nausea is different for the two drugs.

Now, in this experiment each subject must have received the drugs in a certain order, either A first (A, B) or B first (B, A). In most cases in such an experiment an equal number would be given the drugs in order (A, B) as in order (B, A). However, this balance may not survive in the unlike pairs on which the test is based. For example, two possible outcomes for the data of Table 2.8 are shown below:

Outcome 1

| | Order of drug | | |
	(A, B)	(B, A)	
Nausea with A	7	6	13
Nausea with B	1	2	3
	8	8	16

Outcome 2

| | Order of drug | | |
	(A, B)	(B, A)	
Nausea with A	2	11	13
Nausea with B	3	0	3
	5	11	16

In Outcome 1 the first administered drug produces nausea in 9 cases and the second in 7 cases. In Outcome 2 these values are 2 and 14. In both instances nausea occurs with drug A in 13 cases and with drug B in 3 cases. The first outcome suggests that the drugs differ and that A causes more subjects to suffer from nausea. However, the order of administration appears unimportant. For Outcome 2 the issue of drug differences is not so clear since there appears to be a possible order effect with the second administered drug causing more subjects to suffer from the side-effect. Gart (1969) derives a test for an order effect and a test for a treatment or drug effect. The tests use the pairs of observations giving unlike responses, first arranged as in Table 2.9 and then as in Table 2.10.

The terms y_a, y_a', y_b, y_b', n, and n' are used to represent the following frequencies:

y_a: the number of observations for which drug A produces a positive result (in this case causes nausea) in pairs for which A is given first;

y_a': the number of observations for which drug A produces a positive result in pairs for which B is given first;

TABLE 2.9. Data arranged so as to test for a treatment effect.

	Drug order		
	(A, B)	(B, A)	
Nausea with first drug	y_a	y_b'	$y_a + y_b'$
Nausea with second drug	y_b	y_a'	$y_b + y_a'$
	n	n'	$n + n'$

TABLE 2.10. Data arranged so as to test for an order effect.

	Drug order		
	(A, B)	(B, A)	
Nausea with A	y_a	y_a'	$y_a + y_a'$
Nausea with B	y_b	y_b'	$y_b + y_b'$
	n	n'	$n + n'$

y_b: the number of observations for which drug B produces a positive result in pairs for which A is given first;

y_b': the number of observations for which drug B produces a positive result in pairs for which B is given first;

n: the number of unlike pairs with A first;

n': the number of unlike pairs with B first.

Gart shows that Fisher's exact test applied to Table 2.9 gives a test of the hypothesis that there is no difference between the drugs with respect to incidence of nausea. The same test applied to Table 2.10 is a test of the hypothesis that there is no order effect. To illustrate the use of Gart's test it is now applied to Outcomes 1 and 2 for the sixteen unlike pairs of the drug example.

2.6.1. Numerical example of the application of Gart's test

Outcome 1. Tables 2.11 (a) and (b) show the arrangements of the unlike pairs of observations needed for the application of Gart's procedure. Performing Fisher's test on Table 2.11(a) gives a test of the treatment effect, and on Table 2.11(b) a test of the order effect; the results are as follows:

Treatment effect : $P = 0.02$

Order effect : $P = 0.50$

These results indicate that the incidence of nausea is higher for drug A than for drug B, and that the order of drug administration has no effect on the response.

Outcome 2. For this outcome the application of Fisher's test to

TABLE 2.11.

	(a) Drug order				(b) Drug order		
	(A, B)	(B, A)			(A, B)	(B, A)	
Nausea with 1st drug	$7(y_a)$	$2(y_b')$	9	Nausea with A	7	6	13
Nausea with 2nd drug	$1(y_b)$	$6(y_a')$	7	Nausea with B	1	2	3
	8	8	16		8	8	16

TABLE 2.12.

	(a)				(b)		
	Drug order				Drug order		
	(A, B)	(B, A)			(A, B)	(B, A)	
Nausea with 1st drug	2	0	2	Nausea with A	2	11	13
Nausea with 2nd drug	3	11	14	Nausea with B	3	0	3
	5	11	16		5	11	16

Tables 2.12 (a) and (b) gives the following results:

$$\text{Treatment effect}: P = 0.080$$

$$\text{Order effect} \quad : P = 0.018$$

Therefore in this case we conclude that, whereas the drugs do not differ, there is a greater incidence of nausea on the second occasion of drug administration than on the first.

2.7. Combining information from several 2 × 2 tables

In many studies a number of 2×2 tables, all bearing on the same question, may be available, and we may wish to combine these in some way to make an overall test of the association between the row and column factors. For example, in an investigation into the occurrence of lung cancer among smokers and non-smokers, data may be obtained from several different locations or areas, and for each area the data might be arranged in a 2×2 table. Again, in an investigation of the occurrence of a particular type of psychological problem in boys and girls, data may be obtained from each of several different age groups, or from each of several different schools. The question is how may the information from separate tables be pooled?

One obvious method which springs to mind is to combine all the data into a single 2×2 table for which a chi-square statistic is computed in the usual way. This procedure is legitimate only if corresponding proportions in the various tables are alike. Consequently, if the proportions vary from table to table, or we suspect that they vary, this procedure should not be used, since the combined data will not accurately reflect the information contained in the original tables. For example, in the lung cancer and smoking example mentioned previously, where data are collected from several different areas, it may well be the case that the occurrence of lung cancer is more frequent in some areas than in others. Armitage (1971) gives an extreme example of the tendency of this procedure to create significant results.

Another technique which is often used is to compute the usual chi-square value separately for each table, and then to add them; the resulting statistic may then be compared with the value of chi-square from tables with g degrees of freedom where g is the number of separate tables. (This is based on the fact that the sum of g chi-square variables each with one degree of freedom is itself distributed as chi-square with g degrees of freedom.) This is also a poor method since it takes no account of the direction of the differences between the proportions in the various tables, and consequently lacks power in detecting a difference that shows up consistently in the same direction in all or most of the individual tables.

Techniques that are more suitable for combining the information from several 2 × 2 tables are the $\sqrt{(\chi^2)}$ method and Cochran's method, both of which will now be described.

2.7.1. The $\sqrt{(\chi^2)}$ Method

If the sample sizes of the individual tables do not differ greatly (say by more than a ratio of 2 to 1), and the values taken by the proportions lie between approximately 0.2 and 0.8, then a method based on the sum of the square roots of the χ^2 statistics, taking account of the signs of the differences in the proportions, may be used. It is easy to show that, under the hypothesis that the proportions are equal, the χ value for any of the 2 × 2 tables is approximately normally distributed with mean zero and unit standard deviation, and therefore the sum of the χ values for the complete set of g tables is approximately normally distributed with mean zero and standard deviation $\sqrt{g}$. Therefore, as a test statistic for the hypothesis of no difference in the proportions for all the g tables, we may use the statistic Z given by:

$$Z = \sum_{i=1}^{g} \chi_i / \sqrt{g} \qquad (2.8)$$

where χ_i is the value of the square root of the χ^2 statistic for the ith table, with appropriate sign attached.

To illustrate the method let us consider the data shown in Table 2.13 in which the incidence of malignant and benignant tumours in the left and right hemispheres in the cortex is given. The problem is to test whether there is an association between hemisphere and type of tumour. Data for three sites in each hemisphere were available, but an earlier investigation had shown that there was no reason to suspect that any relationship between hemisphere and type of tumour would differ from one site to another, so an overall assessment of the hemisphere–tumour relationship was indicated. For each of the three sites the numbers of patients (33, 27, and 34 respectively) are roughly equal, so we shall apply the $\sqrt{(\chi^2)}$ method to these data. The value of χ^2 is first computed for each separate table. (Note that none of these is significant.) The square roots of these values are then obtained and the sign of the difference between the proportions is assigned to each value of χ. For these data the difference between the proportions is in the same direction for each of the three tables, namely the proportion of malignant tumours in the right hemisphere is always higher than in the left. Consequently the *same* sign is attached to each χ value. (Whether this is positive or negative is, of course,

TABLE 2.13. Incidence of tumours in the two hemispheres for different sites in the cortex.

Site of tumour	Benignant tumours	Malignant tumours	Proportion of malignant tumours	χ^2	χ
1. Left hemisphere	17	5	0.2273	1.7935	1.3392
Right hemisphere	6	5	0.4545		
	23	10			
2. Left hemisphere	12	3	0.2000	1.5010	1.2288
Right hemisphere	7	5	0.4167		
	19	8			
3. Left hemisphere	11	3	0.2143	2.003	1.4155
Right hemisphere	11	9	0.4500		
	22	12			

immaterial.) The test statistic for the combined results is given by

$$Z = \frac{1.339 + 1.2288 + 1.4155}{\sqrt{3}}$$

$$= 2.300$$

This value is referred to the tables of the standard normal distribution and is found to be significant at the 5% level. Therefore, taking all three sites together suggests that there is an association between type of tumour and hemisphere.

If for these data we were to add the separate chi-square statistics, namely 1.7935, 1.5010, and 2.003, we obtain a value of 5.2975. This would be tested against a chi-square with three degrees of freedom, and is not significant at the 5% level. Since in this case the differences are all in the same direction, the $\sqrt{(\chi^2)}$ method is more powerful than one based on summing individual χ^2 values.

2.7.2. Cochran's Method

If the sample sizes and the proportions do not satisfy the conditions mentioned in Section 2.7.1, then addition of the χ values tends to lose power. Tables that have very small N values cannot be expected to be of as much use as those with large N for detecting a difference in the proportions, yet in the $\sqrt{(\chi^2)}$ method all tables receive the same weight. Where differences in the sample sizes are extreme we need some method of weighting the results from individual tables. Cochran (1954) suggested an alternative test based on a weighted mean of the differences between proportions. The test statistic he suggests is Y given by:

$$Y = \sum_{i=1}^{g} w_i d_i \bigg/ \left(\sum_{i=1}^{g} w_i P_i Q_i \right)^{1/2} \tag{2.9}$$

where g is the number of 2×2 tables and for the ith of these:
   n_{i1} and n_{i2} are the sample sizes in the two groups;
   p_{i1} and p_{i2} are the observed proportions in the two samples;
   $P_i = (n_{i1}p_{i1} + n_{i2}p_{i2})/(n_{i1} + n_{i2})$ and $Q_i = (1 - P_i)$;
   $d_i = (p_{i1} - p_{i2})$ and $w_i = n_{i1} n_{i2}/(n_{i1} + n_{i2})$.

It is seen that Y is a weighted mean of the d_i values, in which the weights used give greater importance to differences based on large than on small samples. Under the hypothesis that the d_i terms

TABLE 2.14. The Incidence of tics in three samples of maladjusted children.

Age range		Tics	No tics	Total	Proportion with tics
5–9	Boys	13	57	70	0.1857
	Girls	3	23	26	0.1154
	Total	16	80	96	0.1667
10–12	Boys	26	56	82	0.3171
	Girls	11	29	40	0.2750
	Total	37	85	122	0.3033
13–15	Boys	15	56	71	0.2113
	Girls	2	27	29	0.0690
	Total	17	83	100	0.1700

are zero for $i = 1, \ldots, g$, that is for all tables, then the statistic Y is distributed normally with zero mean and unit variance.

To illustrate Cochran's procedure we shall apply it to the data shown in Table 2.14 where the incidence of tics in three age groups of boys and girls is given. For these data we have three 2×2 tables; hence $g = 3$ and the various quantities needed to perform Cochran's test are as follows:

(a) Age range 5–9 $\quad n_{11} = 70 \qquad n_{12} = 26$
$$p_{11} = 0.1857 \quad p_{12} = 0.1154$$
$$P_1 = 0.1667 \quad Q_1 = 0.8333$$
$$d_1 = 0.0703 \quad w_1 = 18.96$$

(b) Age range 10–12 $\quad n_{21} = 82 \qquad n_{22} = 40$
$$p_{21} = 0.3171 \quad p_{22} = 0.2750$$
$$P_2 = 0.3033 \quad Q_2 = 0.6967$$
$$d_2 = 0.0421 \quad w_2 = 26.89$$

(c) Age range 13–15 $\quad n_{31} = 71 \qquad n_{32} = 29$
$$p_{31} = 0.2113 \quad p_{32} = 0.0690$$
$$P_3 = 0.1700 \quad Q_3 = 0.8300$$
$$d_3 = 0.1423 \quad w_3 = 20.59$$

Application of formula (2.9) gives:

$$Y = \frac{18.96 \times 0.0703 + 26.89 \times 0.0421 + 20.59 \times 0.1423}{(18.96 \times 0.1667 \times 0.8333 + 26.89 \times 0.3033 \times 0.6967 + 20.59 \times 0.1700 \times 0.8300)^{1/2}}$$

$$= 1.61$$

Referring this value to a normal curve it is found to correspond to a probability of 0.1074. Had the three age groups been combined and an overall chi-square test been performed, a value of 2.110 would have been obtained. This corresponds to a probability of 0.2838 which is more than twice that given by Cochran's criterion. This fact illustrates the greater sensitivity of Cochran's test.

2.7.3. Further discussion of the $\sqrt{(\chi^2)}$ and Cochran's method of combining 2 × 2 tables

In cases where the relationship between the two variables in the separate 2 × 2 tables is obviously very different, neither the $\sqrt{(\chi^2)}$ nor Cochran's method is likely to be very informative. For example, suppose we had just two 2 × 2 tables with almost the same sample size, which we wished to combine. If the differences in the proportions of interest in the two tables were large, approximately equal in magnitude, but *opposite* in sign, then both the $\sqrt{(\chi^2)}$ and Cochran's statistic would be approximately zero and therefore yield a non-significant result. Investigators should therefore keep in mind that both methods are really for use when we are trying to detect small systematic differences in proportions. Application of these tests to sets of tables in which these differences vary greatly in magnitude and in direction should be avoided. In such cases combination of the tables in any way is not to be recommended, and they are perhaps best dealt with by the methods to be described in Chapters 4 and 5.

An excellent extended account of the problems of combining evidence from fourfold tables is given in Fleiss (1973), Ch. 10.

2.8. Relative risks

So far in this chapter we have confined our attention to tests of the significance of the hypothesis of no association in 2 × 2 tables. However, important questions of *estimation* may also arise where the null hypothesis is discarded. This is particularly true of certain types of study; for example, in studying the aetiology of a disease it is often useful to measure the increased risk (if any) of incurring a particular disease if a certain factor is present. Let us suppose that for such examples the *population* may be enumerated in terms of the entries in Table 2.15.

The entries in Table 2.15 are *proportions* of the total population.

TABLE 2.15.

| | | Disease | | |
		Present $(+)$	Absent $(-)$	
Factor	Present $(+)$	P_1	P_3	$P_1 + P_3$
	Absent $(-)$	P_2	P_4	$P_2 + P_4$
		$P_1 + P_2$	$P_3 + P_4$	1

If the values of these proportions were known, the risk of having the disease present for those individuals having the factor present would be:

$$P_1/(P_1 + P_3) \qquad (2.10)$$

and for those individuals not having the factor present it would be:

$$P_2/(P_2 + P_4) \qquad (2.11)$$

In many situations involving this type of example, the proportion of subjects having the disease will be small; consequently P_1 will be small compared with P_3, and P_2 will be small compared with P_4; the *ratio* of the risks given by (2.10) and (2.11) then becomes very nearly:

$$\frac{P_1 P_4}{P_2 P_3} \qquad (2.12)$$

This is properly known as the *approximate relative risk*, but it is often referred to simply as *relative risk* and denoted by ψ. If we

TABLE 2.16.

| | | Disease | | |
		$+$	$-$	
Factor	$+$	a	b	$a+b$
	$-$	c	d	$c+d$
		$a+c$	$b+d$	N

have the *sample* frequencies shown in Table 2.16 then ψ may be estimated simply by:

$$\hat{\psi} = \frac{ad}{bc} \qquad (2.13)$$

In general we would require not merely a *point estimate* of ψ as given by (2.13) but a *confidence interval* also. (See Mood and Greybill, Ch. 11.) This is most easily achieved by consideration initially of $\log_e \psi$ since its variance may be estimated very simply as follows:

$$\text{var}(\log_e \hat{\psi}) = \frac{1}{a} + \frac{1}{b} + \frac{1}{c} + \frac{1}{d} \qquad (2.14)$$

By assuming normality we can then obtain a confidence interval for $\log_e \hat{\psi}$, namely:

$$\log_e \hat{\psi} \pm 1.96 \times \sqrt{[\text{var}(\log_e \hat{\psi})]} \qquad (2.15)$$

(This would give a 95% confidence interval.) Taking exponentials of the quantities in (2.15) we would arrive finally at the required confidence interval for ψ. An example will help to clarify this procedure.

2.8.1. *Calculating a confidence interval for relative risk*

Suppose that, of the members of a given population equally exposed to a virus infection, a percentage (which we shall assume contains a fair cross-section of the population as a whole) has been inoculated. After the epidemic has passed, a random sample of people from the population is drawn; the numbers of inoculated and uninoculated that have escaped infection are recorded and the figures in Table 2.17 obtained.

It is clear from the data in Table 2.17 that the proportion of uninoculated people that was infected by the virus is considerably

TABLE 2.17. Incidence of virus infection.

	Not infected	Infected	
Not inoculated	130 (d)	20 (c)	150
Inoculated	97 (b)	3 (a)	100
	227	23	250

larger than the proportion of inoculated infected. In other words the risk of being infected had you been inoculated is less than the risk had you not been inoculated. To quantify this difference we will find a confidence interval for ψ the relative risk.

From (2.13) we first calculate $\hat{\psi}$ the estimate of ψ; this gives, for these data, $\hat{\psi} = 0.201$. Therefore $\log_e \hat{\psi}$ is -1.60. From (2.14) we may now estimate the variance of $\log_e \hat{\psi}$, and obtain a value of 0.401. A 95% confidence interval for $\log_e \psi$ is now obtained using (2.15); substituting the values already calculated for $\log_e \hat{\psi}$ and its variance, we arrive at the interval:

$$-1.60 \pm 1.96 \times 0.63$$

that is

$$-2.83 \text{ to } -0.37$$

We may now take exponentials of these two limits to give the required 95% confidence interval for ψ; this leads to the values 0.06 and 0.69. We are now in a position to say that at the 95% confidence limits, that is with a chance of being wrong on the average once in twenty times, the risk that an inoculated person will be affected by the virus is at most 69% of that of an uninoculated person, and it may be as low as 6%.

Frequently an estimate of relative risk is made from each of a number of sub-sets of the data, and we may be interested in combining these various estimates. One approach is to take separate estimates of $\log_e \psi$ and weight them by the reciprocal of their variance [formula (2.14)]. The estimates may then be combined by taking a weighted mean. A rather simpler pooled estimate of relative risk is that due to Mantel and Haenszel (1959). This, using an obvious nomenclature for the various 2×2 tables involved, is given by:

$$\hat{\psi}_{\text{pooled}} = \frac{\Sigma(a_i d_i / N_i)}{\Sigma(b_i c_i / N_i)} \tag{2.16}$$

In practice this gives very similar results to those obtained by the more complicated procedure outlined involving a weighted mean. Examples are given by Armitage (1971), Ch. 16.

2.9. Guarding against biased comparisons

In Chapter 1 the need to use random or representative samples as a safeguard against obtaining biased results in an investigation was stressed. Now that we have a few examples to which to refer, some further discussion of the matter will be helpful.

An important advance in the development of statistical science was achieved when the advantages of *design* in experimentation were realized (Fisher, 1950). These advantages result from conducting an investigation in such a way that environmental effects and other possible disruptive factors, which might make interpretation of the results ambiguous, are kept under control. But in many investigations in social medicine and in survey work in general, where the data are often of a qualitative kind (and chi-square tests are commonly required), planned experiments are difficult to arrange (Taylor and Knowelden, 1957, Ch. 4). One of the problems is that the occurrence of the phenomenon being studied may be infrequent, so the time available for the investigation permits a retrospective study only to be undertaken. With such studies it is generally difficult to get suitable control data, and serious objections often arise to the samples one might draw, because of limitations in the population being sampled. Berkson (1946) has drawn attention to this point where hospital populations are concerned, and he has demonstrated that the subtle differential selection factors which operate in the referral of people to hospital are likely to bias the results of investigations based on samples from these populations. His main point can be illustrated best by an example.

Suppose an investigator wished to compare the incidence of tuberculosis of the lung in postmen and bus drivers. He might proceed by drawing two samples from the entrants to these occupations in a given month or year and do a follow-up study, with regular X-ray examinations, over a period of years to obtain the information he required. He would, of course, be aware of the possibility that people, by reason of their family histories or suspected predispositions to special ailments, might tend to choose one occupation rather than the other, and he might take steps to control for such possibilities and to eliminate other possible sources of bias. But suppose that, since time and the facilities at his disposal did not permit a prospective study to be carried out, he decided to obtain his samples by consulting the files of a large hospital and extracting for comparison all the postmen and bus drivers found there, the data obtained might not give a true picture. For instance, it might be the case that bus drivers, by virtue of the special responsibility attached to their jobs, were more likely than postmen to be referred to hospital should tuberculosis be suspected.

If this were so, a biased comparison would clearly result.

A biased comparison would also result were it the case that bus drivers, say, were prone to be affected by multiple ailments such as bronchitis and tuberculosis, or carcinoma of the lung and bronchitis, or all three, since these ailments would be likely to aggravate each other so that a bus driver might be referred to hospital for bronchial treatment and then be found to have tuberculosis. Were this a common occurrence then a comparison of postmen and bus drivers as regards the incidence of tuberculosis, based on such hospital samples, would not give true reflection of the incidence of the disease in these occupations in the community.

The relevance of the above discussion to the interpretation of results from investigations such as those reported in this chapter can now be examined. For instance, if we return to Table 2.4 it is clear that the chi-square test applied to the data in it yields an unbiased result only in so far as we can be sure that the hospital populations of psychotics and neurotics from which the samples are drawn are not affected by differential selection. In particular we would want to satisfy ourselves that 'suicidal feelings' did not play a primary part in the referral of neurotics to hospital in the first place. If it did the results given by the chi-square test would be biased.

But it is well to add that Berkson (1946) notes certain conditions under which unbiased comparisons can be made between samples drawn from sources in which selective factors are known to operate. For instance, if the samples of postmen and bus drivers drawn from the hospital files are selected according to some other disease or characteristic unrelated to tuberculosis, say those who on entry to hospital were found to require dental treatment, then a comparison of the incidence of tuberculosis in these men would yield an unbiased result.

As a means of avoiding a biased comparison between samples from a biased source it might too be thought that a one-to-one matching of subjects from the populations to be compared would overcome the difficulty. But clearly this could not act as a safeguard. In the example discussed earlier in the chapter, in which the effect of the symptom 'depersonalization' on the prognosis of endogenous depressed patients was assessed, were it the case – which is unlikely – that 'depersonalization' itself was a primary factor in causing depressed patients to come to hospital, then the result of the comparison made in that investigation would be open to doubt.

2.10. Summary

In this chapter the analysis of 2 × 2 contingency tables has been considered in some detail. However, some warning should be given against rushing to compute a chi-square for every 2 × 2 table which the reader may meet. He should first have some grounds for thinking that the hypothesis of independence is of interest before he proceeds to test it. Secondly, calculation of the chi-square statistic is often just a time-filler and a ritual and may prevent him from thinking of the sort of analysis most needed. For example, in many cases of 2 × 2 tables arising from survey data, the need is for a measure of the degree of association rather than a statistical test for association *per se*. Such measures are discussed in the following chapter. In other cases estimation of the relative risk may be what is required.

A detailed mathematical account of some other approaches to the analysis of 2 × 2 tables is available in Cox (1970).

CHAPTER THREE

r × c Contingency tables

3.1. Introduction

The analysis of $r \times c$ contingency tables, when either r or c or both are greater than 2, presents special problems not met in the preceding chapter. For example, the interpretation of the outcome of a chi-square test for independence in the case of a 2×2 table is clear, namely the equality or otherwise of two proportions. In the case of contingency tables having more than one degree of freedom such interpretation is not so clear, and more detailed analyses may be necessary to decide just where in the table any departures from independence arise. Such methods are discussed in this chapter. Again, a problem not encountered with 2×2 tables is variables having *ordered* categories, for example severity of disease, say 'low', 'medium', or 'high', or amount of smoking, 'none', 'less than 10 a day', 'from 11 to 20 a day' or 'more than 20 a day'. The analysis of tables containing such variables presents further special problems, and these are also considered here. We begin, however, with a numerical example of the usual chi-square test of independence applied to a 3×3 table.

TABLE 3.1. Incidence of cerebral tumours.

		Type			
---	---	A	B	C	
	I	23	9	6	38
Site	II	21	4	3	28
	III	34	24	17	75
		78	37	26	141

3.2. Numerical example of chi-square test

Table 3.1 shows a set of data in which 141 individuals with brain tumours have been doubly classified with respect to type and site of tumour. The three types were as follows: A, benign tumours; B, malignant tumours; C, other cerebral tumours. The sites concerned were: I, frontal lobes; II, temporal lobes; III, other cerebral areas.

In this example $r = c = 3$, and our null hypothesis, H_0, is that site and type of tumour are independent. We first compute the expected frequencies assuming H_0, using formula (1.9). These are shown in Table 3.2.

For example, the entry in the first cell of Table 3.2, namely E_{11}, is obtained as:

$$E_{11} = \frac{38 \times 78}{141} = 21.02$$

Similarly

$$E_{12} = \frac{38 \times 37}{141} = 9.97, \, etc.$$

Although frequencies such as 21.02 are obviously not possible, the terms after the decimal point are kept to increase the accuracy in computing χ^2. [Note that the marginal totals of the expected values are equal to the corresponding marginal totals of observed values, that is $E_{i.} = n_{i.}$ for $i = 1, ..., r$, and $E_{.j} = n_{.j}$ for $j = 1, ..., c$. That this must *always* be true is easily seen by summing equation (1.9) over either i or j.]

Using formula (1.13) we obtain:

$$\chi^2 = \frac{(23.0 - 21.02)^2}{21.02} + \frac{(9.0 - 9.97)^2}{9.97} + ... + \frac{(17.0 - 13.83)^2}{13.83}$$

$$= 0.19 + 0.09 + ... + 0.72$$

$$= 7.84$$

TABLE 3.2. Expected frequencies for the data of Table 3.1.

		Type			
		A	B	C	
	I	21.02	9.97	7.01	38
Site	II	15.49	7.35	5.16	28
	III	41.49	19.68	13.83	75
		78	37	26	141

Table 3.1 has 4 d.f. Examining Appendix A at the 5% level gives the value of chi-square as 9.49. Since χ^2 is less than the tabulated value, we accept the null hypothesis that the two classifications are independent; consequently no association between site and type of tumour can be claimed on the evidence obtained from these data.

3.3. Small expected frequencies

The derivation of the chi-square distribution as an approximation for the distribution of the statistic χ^2 is made under the assumption that the expected values are not 'too small'. This vague term has generally been interpreted as meaning that all expected values in the table should be greater than 5 for the chi-square test to be valid. Cochran (1954) has pointed out that this 'rule' is too stringent, and suggests that if relatively few expectations are less than 5 (say, one cell out of five) a minimum expectation of unity is allowable. Even this rule may be too restrictive, since recent work by Lewontin and Felsenstein (1965), Slakter (1966), and others shows that many of the expected values may be as low as unity without affecting the test greatly. Lewontin and Felsenstein give the following conservative rule for tables in which $r = 2$: 'The $2 \times c$ table can be tested by the conventional chi-square criterion if all the expectations are 1 or greater'. These authors point out that even this rule is extremely conservative and in the majority of cases the chi-square criterion may be used for tables with expectations in excess of 0.5 in the smallest cell.

A procedure that has been used almost routinely for many years to overcome the problem of small expected frequencies is the pooling of categories. However, such a procedure may be criticized for several reasons. Firstly, a considerable amount of information may be lost by the combination of categories, and this may detract greatly from the interest and usefulness of the study. Secondly, the randomness of the sample may be affected. The whole rationale for the chi-square test rests on the randomness of the sample, and that the categories into which the observations may fall are chosen in advance. Pooling categories after the data are seen may affect the random nature of the sample, with unknown consequences. Lastly, the manner in which categories are pooled can have an important effect on the inferences one draws. The practice of combining classification categories should therefore be avoided if at all possible.

3.4. Isolating sources of association in r × c tables

A significant overall chi-square test for an r × c contingency table indicates non-independence of the two variables, but provides no information as to whether non-independence occurs throughout or in a specific part of the table. Therefore one would like to make additional comparisons of cells within the whole table. Various methods have been suggested for this purpose and some of these are discussed in this section. (The situation may be thought of as analogous to that arising when using analysis of variance techniques, where, having found that a set of means differ, one wants to identify just which means differ from which others.)

3.4.1. The Lancaster and Irwin method for partitioning r × c tables

Lancaster (1949) and Irwin (1949) have shown that the overall chi-square statistic for a contingency table can always be partitioned into as many components as the table has degree of freedom. Each component chi-square value corresponds to a particular 2 × 2 table arising from the original table, and each component is independent of the others. Consequently a detailed examination for departures from independence can be made, enabling those categories responsible for a significant overall chi-square value to be identified. To illustrate the method let us examine the following results regarding the incidence of 'retarded activity' in three samples of patients. The data are shown in Table 3.3

An overall test for association on these data gives a value of 5.70 for χ^2, which with two degrees of freedom just falls short of the 5% level of significance. An examination of the data, however, suggests that though the incidence of the symptom for the first two groups is very alike it occurs more frequently amongst these groups than

TABLE 3.3. Retarded activity amongst psychiatric patients.

	Affective disorders	Schizo-phrenics	Neurotics	Total
Retarded activity	12	13	5	30
No retarded activity	18	17	25	60
	30	30	30	90

in the neurotic group. One might be tempted to combine the affective disorders and the schizophrenics and do a chi-square test on the resulting 2×2 table. Indeed, if this is done a value of χ^2 equal to 5.625 is obtained, which with one degree of freedom is significant just beyond the 2.5% level. But such a procedure, carried out purely in the hope of achieving a significant result, after the overall chi-square test has failed to yield one, would be quite unjustified and contrary to good statistical practice. In Table 3.3 the expected frequencies are all 10 or more, so there is no justification for combining groups on this pretext. Of course, had we decided to combine the first two groups before examining the data, everything would have been in order, though we would by this process have lost one degree of freedom unnecessarily. Methods of partitioning the overall chi-square value provide us with means of examining our data in greater detail and of obtaining more sensitive tests of association than we could have obtained otherwise. Kimball (1954) has supplied convenient formulae for obtaining the chi-square values corresponding to the partitioning method given by Lancaster and Irwin. To introduce them let us consider Table 3.4, a contingency table having r rows and c columns. (This nomenclature is used so that Kimball's formulae may be illustrated more clearly.)

The value of χ^2 computed in the ordinary way from this table has $(r - 1)(c - 1)$ degrees of freedom, and the first step in partitioning is to construct the $(r - 1)(c - 1)$ four-fold tables from which the components of chi-square are calculated. In the case of a 2×3 table, for example, the two four-fold tables may be constructed as follows:

a_1	a_2
b_1	b_2

$(a_1 + a_2)$	a_3
$(b_1 + b_2)$	b_3

For instance, using the data in Table 3.3, the corresponding four-

TABLE 3.4.

$a_1 a_2$	·	·	·	a_c	A
$b_1 b_2$	·	·	·	b_c	B
·					·
·		·		·	
$n_1 n_2$	·	·	·	n_c	N

fold tables are:

$$\begin{array}{c|c} 12 & 13 \\ \hline 18 & 17 \end{array} \qquad\qquad \begin{array}{c|c} 25 & 5 \\ \hline 35 & 25 \end{array}$$

At this stage the reader may ask why we combine the first two columns in preference to any other combination. The answer to this question must be supplied by the investigator; he is free to combine the two columns which are likely to be most meaningful in the light of his prior knowledge about the classification categories concerned, but the decision about the columns that are to be combined should always be made before examining the data to be analysed.

Kimball's formulae for partitioning the overall chi-square value in the case of a $2 \times c$ table are obtained by giving t the values $1, 2, \ldots,$ $(c - 1)$ in turn in the general formula:

$$\chi_t^2 = \frac{N^2[b_{t+1}S_t^{(a)} - a_{t+1}S_t^{(b)}]^2}{ABn_{t+1}S_t^{(n)}S_{t+1}^{(n)}} \tag{3.1}$$

where

$$S_t^{(a)} = \sum_{i=1}^{t} a_i, \quad S_t^{(b)} = \sum_{i=1}^{t} b_i, \quad S_t^{(n)} = \sum_{i=1}^{t} n_i$$

and the other symbols are as defined in Table 3.4. Each χ_t^2 value is a one degree of freedom component of the overall χ^2 value, and so:

$$\chi^2 = \chi_1^2 + \chi_2^2 + \ldots + \chi_{c-1}^2 \tag{3.2}$$

We shall illustrate the use of formula (3.1) by applying it to the data of Table 3.3. For these data $c = 3$ and the overall chi-square statistic, χ^2, is partitioned into two components χ_1^2 and χ_2^2 each having one degree of freedom. Substituting the values $t = 1$ and $t = 2$ into formula (3.1) we obtain the following simplified formulae for χ_1^2 and χ_2^2:

$$\chi_1^2 = \frac{N^2(a_1 b_2 - a_2 b_1)^2}{ABn_1 n_2(n_1 + n_2)} \tag{3.3}$$

$$\chi_2^2 = \frac{N^2[b_3(a_1 + a_2) - a_3(b_1 + b_2)]^2}{ABn_3(n_1 + n_2)(n_1 + n_2 + n_3)} \tag{3.4}$$

On substituting the values in Table 3.3 in these formulae we obtain:

$$\chi_1{}^2 = \frac{90^2(12 \times 17 - 13 \times 18)^2}{30 \times 60 \times 30 \times 30 \times 60} = 0.075$$

$$\chi_2{}^2 = \frac{90^2(25 \times 25 - 5 \times 35)^2}{30 \times 60 \times 30 \times 60 \times 90} = 5.625$$

Since we have already mentioned that χ^2 for these data is 5.700, we see that $\chi_1{}^2 + \chi_2{}^2 = \chi^2$. Each component has a single degree of freedom, and whereas the first is not significant the second is significant beyond the 2.5% level. Partitioning of the overall chi-square value, which itself was not significant, has given us a more sensitive test and we are now in a position to say that, whereas the first two groups of patients do not differ where the symptom 'retarded activity' is concerned, the two groups combined differ significantly from the third group.

It is well to note that formula (3.1) differs slightly from the ordinary formula (Chapter 2) for χ^2 for a 2 × 2 contingency table, in so far as it has an additional term in the denominator while instead on N in the numerator it has N^2. The formula, as it stands, contains no correction for continuity but in cases where such a correction is desirable it can be applied in the usual way. When this is done, additivity [that is formula (3.2)] is no longer exact, but the discrepancy between χ^2 and the sum of its component parts in general is negligible.

The general formula for finding the components of the overall chi-square in the case of an ($r \times c$) table where $r > 2$ is also given by Kimball but since it is very cumbersome the reader is referred to the original article should he require it.

3.4.2. Partitioning 2 × c tables into non-independent 2 × 2 tables

The Lancaster and Irwin method described in the preceding section sub-divides the overall chi-square value into independent component parts such that formula (3.2) holds. Many researchers, however, wish to test specific hypotheses about particular sections of a contingency table, which may result in non-independent partitions from it. This is especially true in the case of a 2 × c table in which the c column categories consist of placebo and various treatment groups and the observed responses are dichotomies. Of prime interest are the 2 × 2 tables used to test for independence between the placebo and each of the treatment groups in turn. Since the chi-square values arising from each table are not now independent of each other, it would be unsatisfactory to test them as chi-square variables with one degree of freedom. Such a procedure would have a serious affect on the value of the significance level and could lead to more differences between placebo and treatments being found than the data actually merit. Brunden (1972) shows that a reasonable method is to choose the significance level at which we wish to perform the test, say α, and to compare the chi-square values obtained from each of the $(c - 1)$ 2×2 tables against the one degree of freedom value from tables at the α' level, where:

$$\alpha' = \frac{\alpha}{2(c - 1)} \qquad (3.5)$$

For example, if we were performing tests at the 5% level, that is $\alpha = 0.05$ and $c = 6$, we would compare each χ^2 value with the one degree of freedom value from tables at the 0.005 level. Let us illustrate the procedure with an example in which five drugs for treating depression are to be compared. Six samples of thirty depressed patients are taken and each patient is given one of the five drugs or a placebo; at the end of two weeks each patient is rated as being 'less depressed' or 'same or worse' than before receiving the drugs. The results are shown in Table 3.5.

TABLE 3.5. Treatment of depression

	Placebo	Drug 1	Drug 2	Drug 3	Drug 4	Drug 5	
Improved	8	12	21	15	14	19	89
Same or worse	22	18	9	15	16	11	91
	30	30	30	30	30	30	180

The overall chi-square test for Table 3.5 gives a value of 14.78, which is significant at the 5% level. We now wish to see which drugs differ from the placebo. We do this by forming five separate 2×2 tables and computing χ^2 for each as follows:

(I)		Placebo	Drug 1		
Improved		8	12	20	$\chi^2 = 1.20$
Same or worse		22	18	40	
		30	30	60	

(II)

	Placebo	Drug 2		
Improved	8	21	29	$\chi^2 = 11.28$
Same or worse	22	9	31	
	30	30	60	

(III)

	Placebo	Drug 3		
Improved	8	15	23	$\chi^2 = 3.45$
Same or worse	22	15	37	
	30	30	60	

(IV)

	Placebo	Drug 4		
Improved	8	14	22	$\chi^2 = 2.58$
Same or worse	22	16	38	
	30	30	60	

(V)

	Placebo	Drug 5		
Improved	8	19	27	$\chi^2 = 8.15$
Same or worse	22	11	33	
	30	30	60	

We wish to test at the 5% level, that is at $\alpha = 0.05$. Since, in this case, $c = 6$, then from formula (3.5) $\alpha' = 0.005$, and the corresponding value of chi-square from tables with 1 d.f. is 7.88. Comparing each of the five χ^2 values with the tabulated value we find that only drugs 2 and 5 differ from the placebo, with respect to the change in the symptom depression. Note that here the component chi-square values do *not* sum to the overall chi-square value, namely 14.78.

Rodger (1969) describes other methods whereby specific hypotheses of interest may be tested in $2 \times c$ tables.

3.4.3. *The analysis of residuals*

A further procedure which may be used for identifying the categories responsible for a significant chi-square value is suggested by

Haberman (1973). This involves examination of the *standardized residuals*, e_{ij}, given by:

$$e_{ij} = (n_{ij} - E_{ij})/\sqrt{E_{ij}} \qquad (3.6)$$

where E_{ij} is obtained from formula (1.9) as $n_{i.}n_{.j}/N$. An estimate of the variance of e_{ij} is given by:

$$v_{ij} = (1 - n_{i.}/N)(1 - n_{.j}/N) \qquad (3.7)$$

Thus for each cell in the contingency table we may compute an *adjusted residual*, d_{ij}, where:

$$d_{ij} = e_{ij}/\sqrt{v_{ij}} \qquad (3.8)$$

When the variables forming the contingency table are independent the terms d_{ij} are approximately normally distributed with mean 0 and standard deviation 1. We shall illustrate how the adjusted residuals may be useful by considering the data in Table 3.6, which concern depression and suicidal tendencies in a sample of 500 psychiatric patients.

The expected values are shown in parentheses in Table 3.6. The value of χ^2 for these data is 71.45 which with four degrees of freedom is highly significant. Let us now examine the adjusted residuals given by formula (3.8). We first compute the terms e_{ij} and v_{ij}; these are shown in Tables 3.7 (a) and (b) respectively. We may now obtain the terms d_{ij} and these are shown in Table 3.8. Comparing the absolute values of the entries in Table 3.8 with the 5% standard normal deviate, namely 1.96, we see that many of the adjusted residuals are significant. The pattern is as might be expected; considerably more of the severely depressed patients either attempt

TABLE 3.6. Severity of depression and suicidal intent in a sample of 500 psychiatric patients.

	Not depressed	Moderately depressed	Severely depressed	Total
Attempted suicide	26 (50.13)	39 (33.07)	39 (20.80)	104
Contemplated or threatened suicide	20 (35.67)	27 (23.53)	27 (14.80)	74
Neither	195 (155.20)	93 (102.40)	34 (64.40)	322
Total	241	159	100	500

TABLE 3.7.

(a) Standardized residuals			(b) Variance of standardized residuals		
− 3.41	1.03	3.99	0.41	0.53	0.62
− 2.62	0.72	3.17	0.44	0.58	0.67
3.19	− 0.93	− 3.79	0.18	0.24	0.29

TABLE 3.8. Adjusted residuals

− 5.33	1.41	5.05
− 3.97	0.95	3.87
7.42	− 1.90	− 7.02

or contemplate suicide than those patients who do not suffer from the symptom.

3.5. Combining $r \times c$ tables

A problem considered in the preceding chapter, namely the combination of a number of 2×2 tables, may also arise for larger contingency tables. The investigator may be interested in the association of two qualitative factors with r and c categories respectively, and data may be available for k sub-groups or strata thus forming k separate tables. These strata may simply be different samples, or they may be different age groups, different countries, etc., and again we would like to combine the tables so as to obtain a more sensitive test of the association between the two variables than is given by each table separately. Various methods are available for combining $r \times c$ tables but, as mentioned in the preceding chapter, the combina- of data from different investigations or samples should be considered only when differential effects from one investigation to another can be ruled out.

In some cases it is legitimate to combine the raw data themselves. For example, in an investigation of the symptom 'retarded activity' in psychiatric patients from the diagnostic categories 'affective disorders', 'schizophrenics', and 'neurotics', data may be available from several different age groups of patients. Such a set of data is shown in Table 3.9.

In this example the age groups might be 'young' and 'old' defined

TABLE 3.9. Retarded activity amongst psychiatric patients.

	Affective disorders	Schizophrenics	Neurotics
Age group 1			
Retarded activity	12 (0.400)	13 (0.433)	5 (0.167)
No retarded activity	18	17	25
	—	—	—
	30	30	30
Age group 2			
Retarded activity	17 (0.425)	15 (0.375)	5 (0.125)
No retarded activity	23	25	35
	—	—	—
	40	40	40

TABLE 3.10. Retarded activity amongst psychiatric patients (combined age groups).

	Affective disorders	Schizophrenics	Neurotics
Retarded activity	29	28	10
No retarded activity	41	42	60
	—	—	—
	70	70	70

by some suitable cut-off point. In Table 3.9 the proportion of patients showing the symptom in each diagnostic group for each age-group is given. Since corresponding proportions are very alike, the samples may safely be combined and on overall analysis performed on the combined data shown in Table 3.10.

The corresponding chi-square values are as follows:
Age group 1 $\chi^2 = 5.700$
Age group 2 $\chi^2 = 9.690$
Combined data $\chi^2 = 15.036$
Comparing these with the value of chi-square from tables with 2 d.f., at the 5% significance level, namely 5.99, we see that the analysis of the combined data leads to a more significant result than the analysis of either table separately.

A further method of combining several $r \times c$ tables, which is preferable in many cases to simply combining raw data, is to calculate expected values for each table separately and then to compare the total observed with the total expected frequencies. The usual χ^2 statistic is used but a complication arises since, in this case, it does not follow the chi-square distribution with $(r-1)(c-1)$ d.f. The correct number of degrees of freedom should be somewhat lower, but a convenient correction is not known. Since the effect

TABLE 3.11. Psychiatrists' views on their training; the values in parentheses are expected values.

(I) Psychopharmacology

	Psychiatric hospital	Teaching hospital	Bethlem-Maudsley hospital	
Very good	18 (21.50)	12 (14.33)	13 (7.17)	43
Adequate	10 (13.00)	12 (8.67)	4 (4.33)	26
Inadequate	32 (25.50)	16 (17.00)	3 (8.50)	51
	60	40	20	120

(II) Psychology

	Psychiatric hospital	Teaching hospital	Bethlem-Maudsley hospital	
Very good	28 (31.00)	18 (20.67)	16 (10.33)	62
Adequate	12 (11.50)	8 (7.67)	3 (3.83)	23
Inadequate	20 (17.50)	14 (11.67)	1 (5.83)	35
	60	40	20	120

(III) Epidemiology

	Psychiatric hospital	Teaching hospital	Bethlem-Maudsley hospital	
Very good	10 (17.84)	16 (12.00)	10 (6.16)	36
Adequate	5 (5.45)	3 (3.67)	3 (1.88)	11
Inadequate	40 (31.71)	18 (21.33)	6 (10.95)	64
	55	37	19	111

TABLE 3.12. Combined Data from Table 3.11 (expected values shown in parentheses are obtained from adding those within sciences)

	Psychiatric hospital	Teaching hospital	Bethlem-Maudsley hospital	
Very good	56 (70.34)	46 (47.00)	39 (23.66)	141
Adequate	27 (29.95)	23 (20.01)	10 (10.04)	60
Inadequate	92 (74.71)	48 (50.00)	10 (25.28)	150
	175	117	59	351

is likely to be quite small, the usual test may be used, although this will be rather conservative.

To illustrate this method let us examine the data shown in Table 3.11, which arise from questioning a number of psychiatrists on their views of training they received in three different areas of the basic sciences. (Any psychiatrist who had not received training in a particular area is not included.)

Each table indicates that psychiatrists trained at the Bethlem-Maudsley hospital are more satisfied with their training than those trained elsewhere. However, in this case an overall test of the satisfaction with teaching in the basic sciences may be of interest, and for this purpose Table 3.12 shows the total observed and total expected frequencies, obtained from summing the values in Table 3.11.

From Table 3.12 we obtain $\chi^2 = 26.95$. As a conservative test this may be compared with the tabulated value of chi-square with 4 d.f. At the 1% level this is 13.28, and therefore we can conclude that there is a very significant association between a psychiatrist's satisfaction with his training in the basic sciences and the hospital from which he obtained that training.

Although the methods discussed here can be useful for the combined analysis of a set of $r \times c$ contingency tables in specific cases, it is important to note that such data are, in general, far better dealt with by using the methods to be described in the following chapter and in Chapter 5.

3.6. Ordered tables

Contingency tables formed by variables having classification categories which fall into a natural order, for example, severity of disease, age group, amount of smoking, etc., may be regarded as frequency tables for a sample from a bivariate population, where the scales for the two underlying continuous variables have been divided into r and c categories respectively. Under such an assumption it is possible to quantify the variables by alloting numerical values to the categories and subsequently to use regression techniques to detect linear and higher order trends in the table. In this way *specific* types of departure from independence may be considered, and consequently more sensitive tests may be obtained than by the use of the usual chi-square statistic.

The simplest method for dealing with such data is by assigning arbitrary scores to the categories of each variable, and then using normal regression techniques on these values. In this way the overall chi-square statistic may be partitioned into component parts due to linear, quadratic, and, if required, higher order trends. To illustrate this technique we shall consider the data in Table 3.13, which consists of a sample of 223 boys classified according to age and to whether they were or were not inveterate liars.

An overall chi-square test on the frequencies in Table 3.13 gave a value of 6.691 which with four degrees of freedom is not significant; hence we might conclude that there was no association between age and lying. However, this is an overall test which covers all forms of departure from independence and is consequently insensitive to departures of a specified type. In this case examination of the proportion of inveterate liars in each age group, namely:

0.286   0.367   0.380   0.458   0.568

indicates that the proportions increase steadily with age, and

TABLE 3.13. Boys' ratings on a lie scale.

| | | Age group | | | | |
| | | 5–7 | 8–9 | 10–11 | 12–13 | 14–15 |
Score :		− 2	− 1	0	1	2
Inveterate liars	1	6	18	19	27	25	95
Non-liars	0	15	31	31	32	19	128
		21	49	50	59	44	223

consequently a test specifically designed to detect a trend in these proportions is likely to be more sensitive.

To arrive at such a test we first need to assign numerical values to the classification categories. This has been done in Table 3.13 where the age groups 5–7 to 14–15 have been alloted scores running from − 2 to + 2, and the lie scale has been quantified by alloting the value + 1 to the category 'investerate liar' and the value zero to 'non-liars'. These quantitative values are chosen quite arbitrarily. They are evenly spaced in the present example but they need not be so. For instance, if it were thought that lying was especially associated with puberty and the immediate post-pubertal period, the scores for age might be taken as − 2, − 1, 0, 3, 6, or some other values which gave greater weight to the age-groups in which we were especially interested. The reason for choosing − 2 as the first score has no significance than that it helps to keep the arithmetic simple since the five scores − 2 to + 2 add to zero.

Having quantified our data we may now proceed to treat them as we would data for a bivariate frequency table and calculate the correlation between the two variates 'age' and 'lying', or the regression of one of the variates on the other. For example, suppose, for the data of Table 3.13, we wished to find the linear regression coefficient of lying (y) and age (x). The formula for estimating the coefficient is:

$$b_{yx} = \frac{C_{yx}}{C_{xx}} \tag{3.9}$$

This has variance given by:

$$V(b_{yx}) = \frac{1}{N} \times \frac{C_{yy}}{C_{xx}} \tag{3.10}$$

where

$$C_{yx} = \sum_{i=1}^{2} \sum_{j=1}^{5} n_{ij} y_i x_j - \left(\sum_{i=1}^{2} n_i. y_i \right)\left(\sum_{j=1}^{5} n_{.j} x_j \right) \Big/ N$$

$$C_{xx} = \sum_{j=1}^{5} n_{.j} x_j^2 - \left(\sum_{j=1}^{5} n_{.j} x_j \right)^2 \Big/ N$$

$$C_{yy} = \sum_{i=1}^{2} n_i. y_i^2 - \left(\sum_{i=1}^{2} n_i. y_i \right)^2 \Big/ N$$

To evaluate the expressions (3.9) and (3.10) it is convenient to draw up the frequency tables shown for the lie scores (y_i), the age scores (x_j), and the products $(y_i x_j)$.

Frequency table for age scores					Frequency table for lie scores				
x_j	x_j^2	$n_{.j}$	$n_{.j}x_j$	$n_{.j}x_j^2$	y_i	y_i^2	$n_{i.}$	$n_{i.}y_i$	$n_{i.}y_i^2$
-2	4	21	-42	84	1	1	95	95	95
-1	1	49	-49	49	0	0	128	0	0
0	0	50	0	0					
1	1	59	59	59					
2	4	44	88	176					
		223	56	368			223	95	95

Frequency table for joint lie, Age Scores

$y_i x_j$	n_{ij}	$n_{ij}y_i x_j$
-2	6	-12
-1	18	-18
0	$(15 + 31 + 31 + 32 + 19 + 19) = 147$	0
1	27	27
2	25	50
	223	47

From these tables we may now calculate C_{yx}, C_{xx}, and C_{yy}:

$$C_{yx} = 47 - 95 \times 56/223$$
$$= 23.14$$

$$C_{xx} = 368 - 56^2/223$$
$$= 353.94$$

$$C_{yy} = 95 - 95^2/223$$
$$= 54.53$$

Substituting these values in (3.9) and (3.10) we obtain:

$$b_{yx} = 23.14/353.94$$
$$= 0.06538$$

$$V(b_{yx}) = 54.53/(223 \times 353.94)$$
$$= 0.0006909$$

The component of chi-square due to linear trend is given by $b_{yx}^2/V(b_{yx})$, that is 6.188713. This has a single degree of freedom; recalling that the overall chi-square value for these data is 6.691 based on four degrees of freedom, the following table may now be drawn up.

Source of variation	d.f.	χ^2
Due to linear regression of lying on age	1	6.189 $p < 0.025$
Departure from regression (obtained by subtraction)	3	0.502 Non-significant
		6.691

It is seen then that, though the overall χ^2 statistic with four degrees of freedom is not significant, the χ^2 value due to regression, based on only one degree of freedom, is significant beyond the 2.5% level. Partitioning the overall value has greatly increased the sensitivity of the test and, returning to the data in Table 3.13, we conclude that there is a significant increase in lying with increase in age for the age range in question. We can further say that the increase is linear rather than curvilinear since departure from linear regression is represented by a chi-square value of only 0.502, based on three degrees of freedom, which is a long way from being significant. (An interesting fact, pointed out by Yates, is that the partition of chi-square obtained above is the same whether the regression coefficient of y on x or that of x on y is used.)

The above discussion has been in terms of testing for trend in a $2 \times c$ table. However, the method mentioned is also applicable to the general contingency table with more than two rows, when both classifications are ordered. Bhapkar (1968) gives alternative methods for testing for trends in contingency tables, which in most cases will differ little from the method discussed above. An investigation of the power of chi-square tests for linear trends has been made by Chapman and Nam (1968).

An alternative to assigning *arbitrary* scores to the classification categories is suggested by Williams (1952), and discussed further by

Maxwell (1974). This involves assigning scores in such a way as to maximize the correlation between the two variates. The details of the technique are outside the scope of this text, but in many cases the scores found are more useful and more informative than those fixed arbitrarily by the investigator.

Ordered contingency tables are discussed further in Chapter 5.

3.7. Measures of association for contingency tables

In many cases when dealing with contingency tables a researcher may be interested in *indexing* the strength of the association between the two qualitative variables involved, rather than in simply questioning its significance by means of the chi-square test. His purpose may be to compare the degree of association in different tables or to compare the results with others previously obtained.

Many measures of association for contingency tables have been suggested, none of which appears to be completely satisfactory. Several of these measures are based upon the χ^2 statistic, which cannot itself be used since its value depends on N, the sample size; consequently it may not be comparable for different tables. A further series of measures, suggested by Goodman and Kruskal (1954), arise from considering the *predictive ability* of one of the variables for the other. Other measures of association have been specifically designed for tables having variables with ordered categories. In this section a brief description only is given of some of the suggested measures, beginning with those based on the χ^2 statistic. A *detailed* account of measures of association *and* their properties such as sampling variance etc. is available in Kendall and Stuart (Vol. 2) and in the series of papers by Goodman and Kruskal (1954, 1959, 1963, 1972).

3.7.1. *Measures of association based on the χ^2 statistic*

Several traditional measures of association are based upon the standard chi-square statistic, which itself is not a convenient measure since its magnitude depends on N, and increases with increasing N. The simplest way to overcome this is to divide the value of χ^2 by N to give what is generally known as the *mean square contingency coefficient*, denoted by ϕ^2:

$$\phi^2 = \chi^2/N \qquad (3.11)$$

However, most researchers are generally happier with measures of association that range between -1 and $+1$ (analogous to the correlation coefficient) or between 0 and 1, with zero indicating independence and unity 'complete association'; consequently ϕ^2 is not very satisfactory since it does not necessarily have an upper limit of 1.

A variation of this measure, suggested by Pearson (1904) and called the *coefficient of contingency*, is given by:

$$P = \sqrt{\left(\frac{\chi^2/N}{1 + \chi^2/N}\right)} \tag{3.12}$$

This coefficient clearly lies between 0 and 1 as required, and attains its lower limit in the case of complete independence, that is when $\chi^2 = 0$. In general, however, P cannot attain its upper limit, and Kendall and Stuart (Vol. 2, Ch. 33) show that, even in the case of complete association, the value of P depends on the number of rows and columns in the table. To remedy this the following function of χ^2 has been suggested:

$$T = \frac{\chi^2/N}{\sqrt{[(r-1)(c-1)]}} \tag{3.13}$$

This again takes the value 0 in the case of complete independence, and, as shown by Kendall and Stuart, may attain a value of $+1$ in the case of complete association when $r = c$ but cannot do so if $r \neq c$. A further modification suggested by Cramer (1946), which may attain the value $+1$ for all values of r and c in the case of complete association, is as follows:

$$C = \frac{\chi^2/N}{\min(r-1, c-1)} \tag{3.14}$$

When the table is square, that is when $r = c$, then $C = T$, but otherwise $C > T$ although the difference between them will not be large unless r and c are very different.

The standard errors of all these coefficients can be deduced from the standard error of the χ^2 statistic, and are given in Kendall and Stuart (Vol. 2, Ch. 33).

The major problem with all the above measures of association is that they have no obvious probabilistic interpretation in the same way as has, for example, the correlation coefficient. Interpretation of obtained values is therefore difficult. This has led Goodman and Kruskal (*op. cit.*) to suggest a number of coefficients that are readily interpretable in a *predictive* sense, and these are now briefly described.

3.7.2. *Goodman and Kruskal's lambda measures*

Goodman and Kruskal in their 1954 paper describe several measures of association that are useful in the situation where the two variables involved cannot be assumed to have any relevant underlying continua, and where there is no natural ordering of interest. The rationale behind these measures is the question: "How much does a knowledge of the classification of one of the variables improve one's ability to predict the classification on the other variable?". Now an investigator might claim that he was interested simply in the 'relationship' between the variables and not in predicting one from another. However, it is difficult to discuss the meaning of an association between the two variables without discussing the degree to which one is predictable from the other and the accuracy of the prediction. It seems reasonable therefore to incorporate this notion of predictability into the formal requirements of any index that purports to measure the degree of association between two variables.

To introduce the measures suggested by Goodman and Kruskal we shall examine the data in Table 3.14 which shows the frequencies obtained when 284 consecutive admissions to a psychiatric hospital are classified with respect to social class and diagnosis. Let us suppose that for these data we are interested in the extent to which knowledge of a patient's social class is useful in predicting his diagnostic category.

First, suppose that a patient is selected at random and you are

TABLE 3.14. Social class and diagnostic category for a sample of psychiatric patients.

| | | Diagnosis (Variable B) | | | | |
		Neurotic	Depressed	Personality disorder	Schizo phrenic	
Social class	1	45	25	21	18	109
(Variable A)	2	10	45	24	22	101
	3	17	21	18	18	74
		72	91	63	58	284

asked to guess his diagnosis knowing nothing about his social class. On the basis of the marginal totals for diagnosis in Table 3.14, the best guess would be 'depressed' since this is the diagnosis with the largest marginal total; consequently the probability of the guess being in error would be given by:

P_1 = P(error in guessing diagnosis when social class is unknown)
= P(patient is neurotic) + P(patient has personality disorder) + P(patient is schizophrenic)
= $1 - P$(patient is depressed)

and therefore

$$P_1 = 1 - 91/284 = 0.68$$

Now suppose that a patient is again selected at random and again you are asked to guess his diagnosis, but in this case you are told his social class. Here the best guess would be the diagnostic category with the largest frequency in the particular social class involved, that is for social class 1 neurotic, and for social classes 2 and 3 depressed. The probability of the guess being in error for each of the three social classes may be obtained as follows:

p_1 = P(error in guessing diagnosis when told patient is social class 1)
= $1 - P$(neurotic in social class 1)
= $1 - 45/109 = 0.59$

p_2 = P(error in guessing diagnosis when told patient is social class 2)
= $1 - P$(depressed in social class 2)
= $1 - 45/101 = 0.55$

p_3 = P(error in guessing diagnosis when told patient is social class 3)
= $1 - P$(depressed in social class 2)
= $1 - 21/74 = 0.72$

The overall probability of an error in guessing the diagnosis of a patient when told his social class may now be obtained as follows:

P_2 = P(error in guessing diagnosis when told social class)
= $p_1 P$(patient is from social class 1) + $p_2 P$(patient is from social class 2) + $p_3 P$(patient is from social class 3)

$$= 0.59 \times \frac{109}{284} + 0.55 \times \frac{101}{284} + 0.72 \times \frac{74}{284} = 0.61$$

We see therefore that the knowledge of a patient's social class reduces to some degree the probability of an error in predicting his diagnostic category. Goodman and Kruskal's index of predictive ability, λ_B, is computed from these probabilities as follows:

$$\lambda_B = \frac{P_1 - P_2}{P_1} \tag{3.15}$$

$$= \frac{0.68 - 0.61}{0.68} = 0.103$$

λ_B is the relative decrease in the probability of an error in guessing diagnosis as between social class unknown and known. As such it is really interpretable. In this example, for instance, we can say that, in prediction of diagnosis from social class, information about social class reduces the probability of error by some 10% on average, an amount that would be unlikely to have any practical significance.

In general λ_B may be calculated as follows:

$$\lambda_B = \frac{\sum\limits_{i=1}^{r} \max\limits_{j} (n_{ij}) - \max\limits_{j} (n_{.j})}{N - \max\limits_{j} (n_{.j})} \tag{3.16}$$

It is, of course, entirely possible to reverse the roles of variables A and B and obtain the index λ_A, which is suitable for predictions of A from B. λ_A would be calculated as follows:

$$\lambda_A = \frac{\sum\limits_{j=1}^{c} \max\limits_{i} (n_{ij}) - \max\limits_{i} (n_{i.})}{N - \max\limits_{i} (n_{.i})} \tag{3.17}$$

In general the two indices λ_B and λ_A will be different since situations may arise where B is quite predictable from A, but not A from B.

The value of λ_B (and, of course, λ_A) ranges between 0 and 1. If the information about the predictor variable does not reduce the probability of making an error in guessing the category of the other variable, the index is zero, and we may conclude that there is no *predictive* association between the two variables. On the other hand, if the index is unity no error is made, given knowledge of the predictor variable, and consequently there is complete predictive association.

The coefficients λ_B and λ_A are specifically designed for the *asymmetric* situation in which explanatory and dependent variables are clearly defined. The same 'reduction in error' approach can be used to produce a coefficient for the symmetric situation where neither variable is specially designated as that to be predicted. Instead we suppose that sometimes one and sometimes the other variable is given beforehand and we must predict the one not given. This coefficient, λ, is given by:

$$\lambda = \frac{\sum\limits_{i=1}^{r} \max_j (n_{ij}) + \sum\limits_{j=1}^{c} \max_i (n_{ij}) - \max_j (n_{.j}) - \max_i (n_{i.})}{2N - \max_j (n_{.j}) - \max_i (n_{i.})} \qquad (3.18)$$

For the data of Table 3.14 we obtain:

$$\lambda = \frac{(45 + 45 + 21) + (45 + 45 + 24 + 22) - 91 - 109}{2.284 - 91 - 109}$$

$$= 0.128$$

This coefficient shows the relative reduction in the probability of an error in guessing the category of either variable as between knowing and not knowing the category of the other. λ will always take a value between that of λ_B and λ_A.

A problem arises in the use of the lambda measures of association when the marginal distributions are far from being uniform. In such cases the values of the indices may be misleadingly low, and with extremely skewed marginal distributions it would appear that any lambda-type measure applied to the raw data may be inappropriate. Other problems associated with these measures are discussed in the papers of Goodman and Kruskal previously referenced.

3.7.3. Association measures for tables with ordered categories

In this section we shall discuss measures of association that are specifically designed for the situation where the variables forming the contingency table have ordered categories. Such measures will take positive values when 'high' values of one variable tend to occur with 'high' values of the other variable, and 'low' with 'low'. In the reverse situation the coefficients will be negative. One obvious method for obtaining a measure of association for ordered tables would be to assign scores to the categories and then compute the product–moment correlation coefficient between the two variables. A difficulty arises, however, in deciding on the appropriate scoring system to use. Many investigators would be unhappy about imposing a metric on the categories in their table, and consequently require a measure of association which does not depend on imposing a set of arbitrary scores. Here three such measures will be discussed, namely the tau statistics of Kendall, Goodman and Kruskal's gamma, and Somers's d.

Kendall's tau statistics   Kendall's tau (τ) is well known as a measure of correlation between two sets of rankings. It may be adapted for the general $r \times c$ contingency table having ordered categories by regarding the table as a way of displaying the ranking of the N individuals according to two variables, for one of which only r separate ranks are distinguished and for the other of which only c separate ranks are distinguished. From this point of view the marginal frequencies in the table are the number of observations 'tied' at the different rank values distinguished. Kendall's tau measures are all based on S which is given by:

$$S = P - Q \qquad (3.19)$$

where P is the number of *concordant* pairs of observations, that is pairs of observations such that their rankings on the two variables are in the same direction, and Q is the number of *discordant* pairs for which rankings on the two variables are in the reverse direction. (For computation of S see later). To obtain a measure of association from S it must be standardized to lie in the range from -1 to $+1$. Different methods of standardization give rise to three different tau statistics:

tau-(a) $$\tau_a = \frac{2S}{N(N-1)} \qquad (3.20)$$

tau-(b) $$\tau_b = \frac{2S}{\sqrt{[(P+Q+X_0)(P+Q+Y_0)]}} \qquad (3.21)$$

where X_0 represents the number of observations tied on the first variable only, and Y_0 the number of observations tied on the second variable only. (Again for computation see later.)

tau-(c) $$\tau_c = \frac{2mS}{N^2(m-1)} \qquad (3.22)$$

where $m = \min(r, c)$.

Tau-(a) is the commonly used measure of rank correlation. It is not applicable to contingency table data since it assumes that there are not tied observations. The other two coefficients may, however, be used to measure association in ordered tables. Kendall and Stuart (Vol. 2, Ch. 33) show that tau-(b) may only attain the values ± 1 for a square table, but that tau-(c) can reach these extreme values apart from a small effect produced when N is not a multiple of m. The main problem with these two measures is that they have no obvious probabilistic interpretation, and consequently the meaning of a value of τ_b of 0.7 or τ_c of 0.6, say, cannot be expressed in words in terms of probabilities or errors in prediction.

Goodman and Kruskal's gamma   Goodman and Kruskal (*op. cit.*) suggest a measure of association for ordered tables also based on S, and given by:

$$\gamma = \frac{S}{P + Q} \qquad (3.23)$$

This coefficient has the considerable advantage of having a direct probabilistic interpretation, namely as the difference in probability of like rather than unlike orders for the two variables when two individuals are chosen at random. γ takes the value $+1$ when the data are concentrated in the upper-left to lower-right diagonal (assuming that both variables are ordered in the same direction either both 'low' to 'high' or both 'high' to 'low'). It takes the value zero in the case of independence, but the converse need *not* hold.

Somers's d   Somers (1962) gives a measure of association for contingency tables with ordered categories which is suitable for the asymmetric case in which we have an explanatory and a dependent variable. This coefficient, d_{yx}, is given by:

$$d_{yx} = \frac{S}{P + Q + Y_0} \qquad (3.24)$$

where x indicates the explanatory and y the dependent variable. In this case Y_0 represents the number of observations tied on the dependent variable. This coefficient has a similar interpretation to that of γ. By noting the relationship:

$$\tau_b^{\,2} = d_{yx} d_{xy} \qquad (3.25)$$

we see that the d's bear the same relationship to Kendall's correlation analogue as the classical regression coefficients bear to the product moment correlation coefficient, namely $r^2 = b_{yx} b_{xy}$. Somers's d

coefficients may therefore be thought of as analogous to the ordinary regression coefficients.

Numerical example illustrating the computation of the tau-statistics, γ, and d. The data in Table 3.15 were collected during an investigation into attempted suicides, and show suicidal intent and a depression rating score for a sample of 91 cases. The data are taken from Birtchnell and Alarcon (1971).

We begin by computing P and Q as follows:

P: Each cell in the table is taken in turn and the number of observations in the cell is multiplied by the number of observations in each cell to its south-east and the terms are summed. Cells in the same row and column are ignored. For the data of Table 3.15 then we have:

$$P = 10(4 + 7 + 2 + 9 + 11 + 17) + 14(7 + 2 + 11 + 17)$$
$$+ 8(2 + 17) + 2(9 + 11 + 17) + 4(11 + 17) + 7(17)$$

$$= 1475$$

Q: Each cell is taken in turn and the number of observations in the cell is multiplied by the number of observations in each cell to its south-west. Again cells in the same row and column are ignored. Therefore for Table 3.15 we have:

$$Q = 14(2 + 5) + 8(2 + 4 + 5 + 9) + 2(2 + 4 + 7 + 5 + 9 + 11)$$
$$+ 4(5) + 7(5 + 9) + 2(5 + 9 + 11)$$

$$= 502$$

We now need to obtain X_0 and Y_0, the number of observations tied on depression rating (the 'x' variable) only, and the number tied

TABLE 3.15. Suicidal intent and depression rating score.

| | | \multicolumn{4}{c}{*Depression rating* (x)} | |
		19	21–29	30–39	39	
	Did not want					
Suicidal	to die	10	14	8	2	34
intent (y)	Unsure	2	4	7	2	15
	Wanted to die	5	9	11	17	42
		17	27	26	21	91

only on suicidal intent (the 'y' variable). These may be computed as follows:

$X_0(Y_0)$: Each cell is taken in turn and the number of observations in the cell is multiplied by the number of observations following it in the particular column (row) involved, and the results are summed. For Table 3.15 we have:

$$X_0 = 10(2 + 5) + 2(5) + 14(4 + 9) + 4(9) + 8(7 + 11)$$
$$+ 7(11) + 2(2 + 17) + 2(17)$$
$$= 591$$

Similarly

$$Y_0 = 10(14 + 8 + 2) + 14(8 + 2) + 8(2) + 2(4 + 7 + 2)$$
$$+ 4(7 + 2) + 7(2) + 5(9 + 11 + 17) + 9(11 + 17) + 11(17)$$
$$= 1096$$

Having obtained P, Q, X_0, and Y_0, we may now find the tau statistics, γ and d; they are as follows:

$$\tau_b = \frac{2(1475 - 502)}{\sqrt{[(1475 + 502 + 591)(1475 + 502 + 1096)]}}$$
$$= 0.69$$

$$\tau_c = \frac{2.3(1475 - 502)}{91^2 \times 2}$$
$$= 0.35$$

$$\gamma = \frac{1475 - 502}{1475 + 502}$$
$$= 0.49$$

$$d_{yx} = \frac{1475 - 502}{1475 + 502 + 1096}$$
$$= 0.32$$

We see that the values of the coefficients differ considerably, but all indicate considerable positive association between depression rating score and suicidal intent. In other words, patients having high depression scores tended to be those who expressed the view that they 'wished to die' in their suicide attempt.

Because of its clear probabilistic interpretation the coefficient γ

is perhaps the most useful for measuring association in ordered tables.

In general all measures of association are used as descriptive statistics and, consequently, questions as to their 'significance' are relatively unimportant. However, significance tests for many of them *are* available and are described in the previously referenced works of Kendall and Stuart, and Goodman and Kruskal. Potential users of these measures should remember that most are specifically designed for particular types of situation and that the choice of the appropriate measure depends on consideration of the type of data involved. Computing any measure without regard to its suitability for the particular data set under investigation would obviously not be very sensible.

3.8. Summary

In this chapter the analysis of the general $r \times c$ contingency table has been considered. Many investigators having arrived at a significant overall chi-square value for such a table would proceed no further. In general, however, more detailed investigation of the reasons for the significant association is needed, using the methods described in the previous sections, such as the analysis of residuals. Also tables formed by variables with ordered categories should be investigated for possible trends.

CHAPTER FOUR

Multidimensional tables

4.1. Introduction

Methods for the analysis of contingency tables arising from *two* categorical variables are generally well known even amongst investigators who are not primarily statisticians. However, methods for the analysis of tables arising from *three* or more such variables are less well known, and this chapter seeks to serve as an introduction to the topic. Such tables may arise in many areas. For example, Table 4.1 shows some data concerning classroom behaviour of schoolchildren. Three variables are involved; the first is a teacher's rating of classroom behaviour into *deviant* or *non-deviant*; the second is a risk index based on several items of home condition thought to be related to deviance, for example, overcrowding, large family size, etc.; the categories of this variable are considered *not at risk* for deviant behaviour and considered *at risk* for deviant behaviour. The third variable is an index of the adversity of school conditions based on items such as pupil turnover, number of free school meals, etc.; this is categorized as *low, medium,* or *high.*

TABLE 4.1. Data on classroom behaviour of ten-year-old schoolchildren.

		Adversity of school condition (k)						
		Low		Medium		High		
Risk index (j)		*Not at risk*	*At risk*	*Not at risk*	*At risk*	*Not at risk*	*At risk*	
	Non-deviant	16	7	15	34	5	3	80
Classroom behaviour (i)								
	deviant	1	1	3	8	1	3	17
		17	8	18	42	6	6	97

The analysis of three-dimensional tables poses entirely new conceptual problems as compared with the analysis of two-dimensional tables. However, the extension from tables of three dimensions to those of four or more, whilst increasing the complexity of the analysis, presents no further new problems; consequently the discussion in this chapter will be in terms of the former. Much work has been done on the analysis of multidimensional contingency tables, especially during the past decade. Lewis (1962) gives an excellent review, and a *selection* of the large number of other relevant references are those of Darroch (1962), Birch (1963), Bishop (1969), Fienberg (1970), and Goodman (1968, 1970, 1971). This chapter serves as an introduction to this work, and will form the basis of the more detailed coverage of the topic given in the following chapter when the fitting of models to contingency tables is introduced.

4.2. Nomenclature for three-dimensional tables

The nomenclature used previously for dealing with an $r \times c$ table is easily extended to deal with a three-dimensional $r \times c \times l$ contingency table having r row, c column, and l 'layer' categories. The observed frequency in the ijkth cell of the table is represented by n_{ijk} for $i = 1,2,...,r$, $j = 1,2,...,c$, and $k = 1,2,...,l$. By summing the n_{ijk} over different subscripts various marginal totals may be obtained. For example, summing over all values of both i and j will yield the total for the kth layer category. Similarly we may obtain the totals for the ith row category, and the jth column category, by summing the n_{ijk} over j and k, and over i and k respectively. These totals will be known as *single variable marginals*, and we have:

$$n_{i..} = \sum_{j=1}^{c} \sum_{k=1}^{l} n_{ijk}$$

$$n_{.j.} = \sum_{i=1}^{r} \sum_{k=1}^{l} n_{ijk} \qquad (4.1)$$

$$n_{..k} = \sum_{i=1}^{r} \sum_{j=1}^{c} n_{ijk}$$

For example, for the data shown in Table 4.1 we have:

Number of non-deviant children $= n_{1..} = (16 + 7) + (15 + 34) + (5 + 3) = 80$.

Number of deviant children $= n_{2..} = (1+1) + (3+8) + (1+3)$ $= 17$.

Number of children not at risk $= n_{.1.} = (16+1) + (15+3)$ $+ (5+1) = 41$.

Number of children at risk $= n_{.2.} = (7+1) + (34+8) + (3+3)$ $= 56$.

Number of children at low adversity schools $= n_{..1} = (16+7)$ $+ (1+1) = 25$.

Number of children at medium adversity schools $= n_{..2} =$ $(15+34) + (3+8) = 60$.

Number of children at high adversity schools $= n_{..3} = (5+3)$ $+ (1+3) = 12$.

Summing the n_{ijk} over any single subscript gives what we shall call the *two variable marginal totals*:

$$n_{ij.} = \sum_{k=1}^{l} n_{ijk}$$

$$n_{i\cdot k} = \sum_{j=1}^{c} n_{ijk} \qquad (4.2)$$

$$n_{.jk} = \sum_{i=1}^{r} n_{ijk}$$

For example, Table 4.2 shows the two-variable marginals for the data of Table 4.1, obtained by summing over the third variable, namely adversity of school condition.

Similar tables could be obtained by summing over either of the other two variables. The grand total of the frequencies, namely $n_{...}$ given by:

$$n_{...} = \sum_{i=1}^{r} \sum_{j=1}^{c} \sum_{k=1}^{l} n_{ijk} \qquad (4.3)$$

TABLE 4.2. Some two variable marginal totals from Table 4.1.

| | | Risk index | |
		Not at risk	At risk
	Non-deviant	36	44
Classroom behaviour			
	deviant	5	12

is generally denoted by N. This system of nomenclature is easily generalized to contingency tables involving more than three ways of classification. (A similar nomenclature is used for the population probabilities, p_{ijk}, and the estimated expected values, E_{ijk}.)

4.3. An introduction to the analysis of multidimensional tables

Researchers with data in the form of a multidimensional contingency table may ask why they should not simply attempt its analysis by examining all the two-dimensional tables arrived at by summing over the other variables. Reasons why this would not, in most cases, be an appropriate procedure are not difficult to find. The most compelling is that it can lead to very misleading conclusions being drawn about the data; why this should be so will become clear after we have discussed concepts such as partial and conditional independence in later sections. Here it will suffice to illustrate the problem by means of an example; the data for this are in Table 4.3 (taken from Bishop, *op. cit.*).

Analysing firstly only the data for clinic A we find that the χ^2 statistic is almost zero. Similarly for the data from clinic B, χ^2 is approximately zero. If we were now to collapse Table 4.3 over clinics and compute the χ^2 statistic for the combined data, we arrive at a value of $\chi^2 = 5.26$ which with 1 d.f. is significant beyond the 5% level; consideration only of this table would therefore lead us to conclude erroneously that survival was related to amount of care received. The reasons for spurious results such as this arising will become apparent later. However, this example should make it clear why consideration of all two-dimensional tables is not a sufficient procedure for the analysis of multidimensional tables. (This example should also serve to illustrate our previous warnings with regard to the combination of contingency table data; see Chapters 2 and 3.)

TABLE 4.3. Three-dimensional contingency table relating survival of infants to amount of pre-natal care received in two clinics.

| | | Infants' survival | | | |
| | | Died | | Survived | |
Amount of pre-natal care		Less	More	Less	More
Place where	Clinic A	3	4	176	293
care received	Clinic B	17	2	197	23

The analysis of multidimensional tables presents problems not met for two-dimensional tables, where a single hypothesis, namely that of the independence of the two variables involved, is of interest. In the case of multidimensional tables, more than one hypothesis may be of concern. For example, the investigator may wish to test that some variables are independent of some others, or that a particular variable is independent of the remainder. The simplest hypothesis of interest for a multidimensional table is that of the mutual independence of the variables; in the following section testing such a hypothesis for a three-dimensional table is considered.

4.4. Testing the mutual independence of the variables in a three-way table

The hypothesis of the mutual independence of the variables in a three-dimensional contingency table may be formulated as follows:

$$H_0: \; p_{ijk} = p_{i..} p_{.j.} p_{..k} \tag{4.4}$$

where p_{ijk} represents the probability of an observation occurring in the ijkth cell, and $p_{i..}$, $p_{.j.}$, and $p_{..k}$ are the marginal probabilities of the row, column, and layer variables respectively. This is the three-dimensional equivalent of the hypothesis of independence in a two-way table [see equation (1.6)]. To test this hypothesis we proceed in an exactly analogous manner to that previously described for the two variable case (again see Section 1.5). Firstly we need to calculate estimates of the frequencies to be expected when H_0 is true. Secondly we need to compare these values with the observed frequencies by means of the usual χ^2 statistic. Lastly we compare χ^2 with the tabulated chi-square value having the relevant number of degrees of freedom. In the case of the hypothesis of the mutual independence of the three variables, the expected values may be obtained in a similar way to that used for two-way tables as follows:

$$E_{ijk} = N \hat{p}_{i..} \hat{p}_{.j.} \hat{p}_{..k} \tag{4.5}$$

[cf. equation (1.9)] where $\hat{p}_{i..}$, $\hat{p}_{.j.}$, and $\hat{p}_{..k}$ are estimates of the probabilities $p_{i..}$, $p_{.j.}$, and $p_{..k}$. It is easy to show that the best estimates are based upon the relevant single variable marginal totals, namely:

$$\hat{p}_{i..} = \frac{n_{i..}}{N}, \; \hat{p}_{.j.} = \frac{n_{.j.}}{N}, \; \hat{p}_{..k} = \frac{n_{..k}}{N} \tag{4.6}$$

(Again, as in Chapter 1, these are maximum likelihood estimates.) Substituting these values in (4.5) gives:

$$E_{ijk} = \frac{N n_{i..} . n_{.j.} . n_{..k}}{N \quad N \quad N}$$

$$= \frac{n_{i..} . n_{.j.} . n_{..k}}{N^2} \tag{4.7}$$

Having obtained the expected values using (4.7), we compute:

$$\chi^2 = \sum_{i=1}^{r} \sum_{j=1}^{c} \sum_{k=1}^{l} \frac{(n_{ijk} - E_{ijk})^2}{E_{ijk}} \tag{4.8}$$

To complete the test we now need to know the degrees of freedom of χ^2. In this case, where the hypothesis is that of the mutual independence of the three variables, 'degrees of freedom' takes the value:

$$d.f. = rcl - r - c - l + 2 \tag{4.9}$$

In the case of other hypotheses which might be of interest (see following section) the value for the degrees of freedom will depend upon the particular hypothesis under test. A general procedure for determining degrees of freedom for chi-square tests on multidimensional tables is discussed in Section 4.7.

4.4.1. Numerical example

To illustrate the test of mutual independence we shall apply it to the data of Table 4.1. First we compute the expected values using formula (4.7). For example, the expected value for the non-deviant, not at risk, low adversity cell, namely E_{111}, is given by:

$$E_{111} = \frac{80 \times 41 \times 25}{97 \times 97} = 8.72$$

and the full set of expected values is given in Table 4.4

Application of formula (4.8) gives $\chi^2 = 17.30$, with degrees of freedom, from formula (4.9), given by:

$$d.f. = 3 \times 2 \times 2 - 3 - 2 - 2 + 2$$
$$= 7$$

At the 5% level the tabulated value of chi-square with 7 d.f. is 14.07, and consequently we are led to reject the hypothesis of mutual independence. More detailed examination of the data is now needed

TABLE 4.4. Expected values for the data of Table 4.1 under the hypothesis that the three variables are mutually independent.

Risk index		Adversity of school condition						
		Low		Medium		High		
		Not at risk	At risk	Not at risk	At risk	Not at risk	At risk	
Classroom behaviour	Non-deviant	8.72	11.90	20.92	28.57	4.18	5.71	80
	Deviant	1.85	2.53	4.44	6.07	0.89	1.21	17
		10.57	14.43	25.36	34.64	5.07	6.92	97

to assess which variables cause this hypothesis to be rejected; this is discussed in the next section.

Examining Table 4.4 we see that summing the expected values over any two of the variables gives a total equal to the relevant single variable marginal total of observed values. For example:

$$E_{1..} = 8.72 + 11.90 + 20.92 + 28.57 + 4.18 + 5.71$$
$$= 80 = n_{1..}$$

However, summing these values over any *single* variable does *not* give totals equal to the two-variable marginals of the observed values. For example:

$$E_{11.} = 8.72 + 20.92 + 4.18$$
$$= 33.82 \neq n_{11.}. \quad (n_{11.} = 36)$$

The constraints on the marginal totals of expected values in the case of the hypothesis of mutual independence are such that only their single variable marginals, namely $E_{i..}$, $E_{.j.}$, and $E_{..k}$, are required to equal the corresponding marginals of observed values, namely $n_{i..}$, $n_{.j.}$, and $n_{..k}$. In the case of other hypotheses various other marginal totals of expected values may be so constrained, as we shall see in the following section. Such constraints arise from the form of the maximum likelihood equations from which the expected values (remember, of course, that we are talking of *estimated* expected values; see Chapter 1) are derived; for details see Birch (*op. cit.*).

4.5. Further hypotheses of interest in three-way tables

If the test of mutual independence described in the preceding section gives a non-significant result, then one concludes that further analysis of the table is unnecessary. However, when the test gives a significant result it should not be assumed that there are significant associations between all variables. It might be the case, for example, that an association exists between two of the variables whilst the third is completely independent. In this case hypotheses of *partial independence* would be of interest. Again, situations arise where two of the variables are independent in each level of the third, but each may be associated with this third variable. In other words, the first two variables are *conditionally independent* given the level of the third. Such hypotheses may again be formulated in terms of probabilities, as we shall illustrate by considering the three hypotheses of partial independence that are possible for a three-dimensional table. These are as follows:

$H_0^{(1)}$: $p_{ijk} = p_{i..} \, p_{.jk}$ (row classification independent of column and layer classification)

$H_0^{(2)}$: $p_{ijk} = p_{.j.} \, p_{i.k}$ (column classification independent of row and layer classification)

$H_0^{(3)}$: $p_{ijk} = p_{..k} \, p_{ij.}$ (layer classification independent of row and column classification)

Let us consider the first of these in more detail. This hypothesis states that the probability of an observation occurring in the ijkth cell, that is p_{ijk}, is given by the product of the probability of it falling in the ith category of the row variable, $p_{i..}$, and the probability of its being in the jkth cell of the column $\times$ layer classification, $p_{.jk}$. If the hypothesis is true it implies that the row classification is independent of both the column *and* the layer classification; that is it implies the truth of the following composite hypothesis:

$$p_{ij.} = p_{i..} \, p_{.j.} \quad \text{and} \quad p_{i.k} = p_{i..} \, p_{..k}$$

To test the hypothesis we proceed in exactly the same manner as previously, beginning with the computation of expected values which in this case are given by:

$$E_{ijk} = N \hat{p}_{i..} \, \hat{p}_{.jk} \tag{4.10}$$

with the estimators $\hat{p}_{i..}$ and $\hat{p}_{.jk}$ of the probabilities $p_{i..}$ and $p_{.jk}$ being obtained from the relevant marginal totals as follows:

$$\hat{p}_{i\cdot\cdot} = \frac{n_{i\cdot\cdot}}{N} \text{ (as before)}, \quad \hat{p}_{\cdot jk} = \frac{n_{\cdot jk}}{N} \tag{4.11}$$

In this case the two variable marginal totals, namely $n_{\cdot jk}$, found by summing the observed frequencies over the first variable are needed. [Again the estimates given by (4.11) are maximum likelihood estimates.] Using these probability estimates in (4.10) we obtain:

$$E_{ijk} = N \frac{n_{i\cdot\cdot}}{N} \frac{n_{\cdot jk}}{N}$$

$$= \frac{n_{i\cdot\cdot}\, n_{\cdot jk}}{N} \tag{4.12}$$

The statistic χ^2 is now calculated using (4.8) and for this hypothesis has degrees of freedom given by:

$$\text{d.f.} = clr - cl - r + 1 \tag{4.13}$$

4.5.1. Numerical example

Let us now test hypothesis $H_0^{(1)}$ for the data of Table 4.1. For these data this hypothesis states that classroom behaviour is independent of the school condition as indicated by the adversity index and the home condition as indicated by the risk index. First we calculate the expected values using formula (4.12). For example,

TABLE 4.5. Expected values for the data of Table 4.1. under the hypothesis that classroom behaviour is independent of the other two variables.

| | | Adversity of school condition | | | | | |
| | | Low | | Medium | | High | |
Risk index		Not at risk	At risk	Not at risk	At risk	Not at risk	At risk
	Non-deviant	14.02	6.60	14.85	34.64	4.95	4.85	80
Classroom behaviour								
	Deviant	2.98	1.40	3.15	7.36	1.05	1.05	17
		17	8	18	42	6	6	97

that for the first cell, namely E_{111}, is given by:

$$E_{111} = \frac{80 \times 17}{97} = 14.02$$

and the full set of expected values are shown in Table 4.5.

Using these expected values we obtain $\chi^2 = 6.19$, with degrees of freedom from formula (4.13) of 5. At the 5% significance level the tabulated value of chi-square with 5 d.f. is 11.07, and hence we are led to accept our hypothesis that classroom behaviour is independent of the other two variables. Since we have already shown that the three variables are not mutually independent, this result leads us to conclude that there is a significant association between the school and the home condition as indicated by the indices for these two factors. If we collapse Table 4.1 into the 2×3 two-dimensional contingency table shown in Table 4.6 by summing over the classroom behaviour variable, and perform the usual chi-square calculation, we can see that this is so. (As we shall see in the following chapter, collapsing the table in this way, *after* showing that the hypothesis of partial independence is acceptable, is a legitimate procedure.)

For Table 4.6, $\chi^2 = 10.78$ with 2 d.f., which is significant beyond the 1% level. As might be predicted, fewer children from schools in the low adversity category have home conditions that put them at risk for deviant behaviour than would be expected if the two variables were independent.

Returning to Table 4.5 we see that in this case, in addition to the single variable marginal totals for expected and observed values being equal, the two-variable marginals obtained by summing over

TABLE 4.6. Data of Table 4.1 summed over classroom behaviour; the expected values under the hypothesis of independence are given in parentheses.

		Adversity of school condition			
		Low	Medium	High	
	Not at risk	17 (10.57)	18 (25.36)	6 (5.2)	41
Risk index					
	At risk	8 (14.43)	42 (34.64)	6 (6.93)	56
		25	60	12	97

classroom behaviour are also equal, that is $E_{.jk} = n_{.jk}$. For example:

$$E_{.11} = 14.02 + 2.98 = 17 = n_{.11}$$

Hypotheses of partial independence fix the single variable and one set of two-variable marginal totals of the expected values to be equal to the corresponding totals of the observed values.

The concept of conditional independence, mentioned briefly earlier, will be considered in more detail in Chapter 5, but it is not difficult to see that the data in Table 4.3 illustrate this type of independence, since amount of care and survival are obviously independent in each level of the third variable (place where care received), that is they are independent in clinic A *and* in clinic B.

4.6. Second-order relationships in three-way tables

For multidimensional tables the possibility exists of the presence of a more complex relationship between variables than those considered up to this point. For example, in a three-way table an association between two of the variables may differ in degree or in direction in different categories of the third; consequently a conjoint three-variable relationship would have to be assumed. Such a relationship would be termed *second order* as opposed to the *first order* associations between pairs of variables considered up to this point. Second and higher order associations are best understood in terms of the models to be considered in the following chapter. However, Roy *et al.* (1956) have considered specifically the formulation of the hypothesis of no second-order association between the variables in a three-way table, in terms of the probabilities p_{ijk}. The formulation these workers give is as follows:

$$H_0 : \frac{p_{rcl}p_{ijl}}{p_{icl}p_{rjl}} = \frac{p_{rck}p_{ijk}}{p_{ick}p_{rjk}} \qquad (4.14)$$

$$i = 1,...,r-1 \; ; \; j = 1,...,c-1 \; ; \; k = 1,...,l-1$$

Although it is not difficult to show how (4.14) relates to the hypothesis of no first-order association in a two-dimensional table, details will not be given here. Interested readers should consult Bhapkar and Koch (1968) who show that (4.14) arises naturally when extending the hypothesis of independence between pairs of variables to that of no three-variable association for higher order tables. Essentially the quantity on the left-hand side of (4.14) represents a measure of the association of the first two variables within the lth *category* of the third, and that on the right-hand side the same measure of association between the first two variables within the kth *category* of the third. The hypothesis states then that this measure of association is the same for all categories of the third variable, or, in other words, that the association between variables 1 and 2 does not differ with the level of variable 3. (Of course, since the order of variables is arbitrary, the hypothesis of no second-order association between the three variables implies that the association between any pair of variables is the same at all levels of the remaining variable.)

A complication arises when we consider how to test the hypothesis given in (4.14), since estimates of the cell frequencies to be expected when H_0 is true are not as easily obtainable as for the other hypotheses considered up to this point. In fact such estimates *cannot* be found directly as products of various marginal totals but instead must be obtained *interatively* by a procedure to be described in the following chapter. Having obtained the estimates, however, the test proceeds in the usual fashion with the computation of the statistic χ^2. Detailed discussion of testing for second and higher order relationships in multidimensional tables will be left until the following chapter.

4.7. Degrees of freedom

A convenient method for determining the degrees of freedom of the χ^2 statistic for multidimensional tables is by use of the following general formula:

d.f. = (Number of cells in table $- 1$) $-$ (Number of probabilities estimated from the data for the particular hypothesis being tested.) (4.15)

For example, let us consider the hypothesis of mutual independence in a three-way table. First, the number of cells in an $r \times c \times l$ table is clearly the product rcl. For this hypothesis we need to estimate the probabilities $p_{i..}$, $p_{.j.}$, and $p_{..k}$ for all values of i, j, and k, using equation (4.6). Since probabilities must sum to unity there are $(r - 1)$ row, $(c - 1)$ column, and $(l - 1)$ layer probabilities to estimate, and hence the degrees of freedom in this case will be:

$$\text{d.f.} = rcl - (r - 1) - (c - 1) - (l - 1) - 1$$
$$= rcl - r - c - l + 2$$

Now consider the hypothesis of partial independence discussed in Section 4.5. In this case we again need to estimate the probabilities $p_{i..}$, but in addition we use equation (4.11) to estimate the probabilities $p_{.jk}$ for all values of j and k. As before there are $(r-1)$ row probabilities, and the number of column $\times$ layer probabilities is simply $(cl-1)$. Therefore we arrive at the following value for the degrees of freedom:

$$\begin{aligned} \text{d.f.} &= rcl - (r-1) - (cl-1) - 1 \\ &= rc - r - cl + 1 \end{aligned}$$

4.8. Likelihood ratio criterion

An alternative criterion to the usual χ^2 statistic for comparing observed frequencies with those expected under a particular hypothesis is the *likelihood ratio criterion*, χ_L^2, given by:

$$\chi_L^2 = 2\Sigma \,\text{Observed} \times \log_e (\text{Observed}/\text{Expected}) \qquad (4.16)$$

which like χ^2 has a chi-square distribution when the hypothesis is true. (The degrees of freedom of χ_L^2 are, of course, the same as for χ^2.) Since χ^2 is easily shown to be an approximation to χ_L^2 for large samples, the two statistics will take similar values for many tables. However, Ku and Kullback (1974), Williams (1976), and others show that, in general, χ_L^2 is preferable to χ^2, and consequently it will be the criterion used in the remainder of this text.

4.9. Summary

In this chapter contingency tables arising from more than two categorical variables have been introduced. Some of the new problems arising from this extension have been considered, in particular the increased number of hypotheses that may be of interest, and the existence of second and higher order relationships. More details are given in the references previously cited, particularly those of Lewis, Birch, and Bishop.

The present chapter serves as an introduction to Chapter 5, in which estimation techniques are discussed, and where the fitting of models to contingency table data is introduced.

CHAPTER FIVE

Log-linear models for contingency tables

5.1. Introduction

The previous chapters have dealt almost exclusively with *hypothesis testing* techniques for the analysis of contingency tables. In this chapter a different approach will be considered, namely that of *fitting models* and of *estimating the parameters* in the models. The term model here refers to some 'theory' or conceptual framework about the observations, and the parameters in the model represent the 'effects' that particular variables or combinations of variables have in determining the values taken by the observations. Such an approach is common in many branches of statistics such as regression analysis and the analysis of variance. Most common are *linear models* which postulate that the expected values of the observations are given by a linear combination of a number of parameters. Techniques such as maximum likelihood and least squares may be used to *estimate* the parameters, and estimated parameter values may then be used in identifying which variables are of greatest importance in determining the values taken by the observations.

The models for contingency table data to be discussed in the following section are very similar to those used for quantitative data in, for example, the analysis of variance, and readers not familiar with such models are referred to Hays (1973), Ch. 12. A consequence of this similarity is that many authors have adopted an analysis of variance term, namely *interaction*, as an alternative to the term association, for describing a relationship between the qualitative variables forming a contingency table. Such a terminology will be used throughout the remainder of this text, and we shall therefore speak of *first-order* interactions between *pairs* of variables, *second-order* interactions between *triplets* of variables, and so on.

Models for contingency tables are well described in the articles referred to in the preceding chapter by Birch, Bishop, Fienberg, and Goodman. The most comprehensive account is, however, given in Bishop, Fienberg, and Holland (1975), which should be regarded as the standard reference for the analysis of complex contingency table data. The description in this chapter follows essentially that given by these authors, but in somewhat simplified form. The major advantages obtained from the use of these techniques are, first that they provide a systematic approach to the analysis of complex multidimensional tables, and second that they provide estimates of the magnitude of effects of interest; consequently they allow the relative importance of different effects to be judged.

5.2. Log-linear models

Let us now consider how the type of model used in the analysis of variance of quantitative data can arise for contingency table data. Returning for the moment to two-dimensional tables, the hypothesis of independence, that is of no first-order interaction between the two variables, specifies, as we have seen previously, that:

$$p_{ij} = p_{i.} p_{.j} \qquad (5.1)$$

This relationship specifies a particular structure or model for the data, namely that in the population the probability of an observation falling in the ijth cell of the table is simply the *product* of the marginal probabilities. We now wish to ask how this model could be rearranged so that p_{ij} or some function of it can be expressed as the *sum* of the marginal probabilities or some function of them? The model would then begin to correspond to those found in the analysis of variance. By taking the natural logarithms of (5.1) such a relationship is easily found, namely:

$$\log_e p_{ij} = \log_e p_{i.} + \log_e p_{.j} \qquad (5.2)$$

This may be rewritten in terms of the theoretical frequencies, $F_{ij}(F_{ij} = Np_{ij}$ etc.; see Chapter 1), as:

$$\log_e F_{ij} = \log_e F_{i.} + \log_e F_{.j} - \log_e N \qquad (5.3)$$

Summing (5.3) over i we have:

$$\sum_{i=1}^{r} \log_e F_{ij} = \sum_{i=1}^{r} \log_e F_{i.} + r\log_e F_{.j} - r\log_e N \qquad (5.4)$$

and over j:

$$\sum_{j=1}^{c} \log_e F_{ij} = c \log_e F_{i\cdot} + \sum_{j=1}^{c} \log_e F_{\cdot j} - c \log_e N \qquad (5.5)$$

and finally over i *and* j we have:

$$\sum_{i=1}^{r} \sum_{j=1}^{c} \log_e F_{ij} = c \sum_{i=1}^{r} \log_e F_{i\cdot} + r \sum_{j=1}^{c} \log_e F_{\cdot j} - rc \log_e N$$

It is now a matter of simple algebra to show that equation (5.3) may be rewritten in a form reminiscent of the models used in the analysis of variance, namely:

$$\log_e F_{ij} = u + u_{1(i)} + u_{2(j)} \qquad (5.7)$$

where

$$u = \frac{\displaystyle\sum_{i=1}^{r} \sum_{j=1}^{c} \log_e F_{ij}}{rc} \qquad (5.8)$$

$$u_{1(i)} = \frac{\displaystyle\sum_{j=1}^{c} \log_e F_{ij}}{c} - \frac{\displaystyle\sum_{i=1}^{r} \sum_{j=1}^{c} \log_e F_{ij}}{rc} \qquad (5.9)$$

$$u_{2(j)} = \frac{\displaystyle\sum_{i=1}^{r} \log_e F_{ij}}{r} - \frac{\displaystyle\sum_{i=1}^{r} \sum_{j=1}^{c} \log_e F_{ij}}{rc} \qquad (5.10)$$

We see that (5.7) specifies a linear model for the logarithms of the frequencies or, in other words, what is generally known as a *log-linear* model. Its similarity to the models used in the analysis of variance is clear; consequently analysis of variance terms are used for the parameters and u is said to represent an 'overall mean effect', $u_{1(i)}$ represents the 'main effect' of the ith category of variable 1, and $u_{2(j)}$ the 'main effect' of the jth category of variable 2. Examining equations (5.9) and (5.10) we see that the main effect parameters are measured as deviations of row or column means of log-frequencies from the overall mean; consequently:

$$\sum_{i=1}^{r} u_{1(i)} = 0, \qquad\qquad \sum_{j=1}^{c} u_{2(j)} = 0$$

or using an obvious dot notation:

$$u_{1(\cdot)} = 0, \qquad\qquad u_{2(\cdot)} = 0$$

The above derivation of the model specified in (5.7) has been in terms of the theoretical frequencies, F_{ij}. In practice, of course, we shall need to estimate these *and* the parameters in the model, and subsequently test the adequacy of the suggested model for the observed data. Details of fitting log-linear models to contingency table data are given in Section 5.3.

We shall also see in Section 5.3 that the values taken by the 'main effect' parameters simply reflect differences between the row or the column marginal totals, and so, in the context of contingency table analysis, are of little concern (in contrast to the analysis of variance situation where main effects are usually of major importance). What is of interest is to extend the model specified in (5.7) to the situation in which the variables are *not* independent. To do this we need to introduce an extra term to represent the interaction between the two variables, giving:

$$\log_e F_{ij} = u + u_{1(i)} + u_{2(j)} + u_{12(ij)} \qquad (5.11)$$

The reason for the nomenclature used for the parameters now becomes clear. The numerical subscripts of the parameters denote the particular variables involved, and the alphabetic subscripts the categories of these variables in the same order. Thus $u_{12(ij)}$ represents the interaction effect between levels i and j of variables 1 and 2 respectively. As we shall see in the following section, interaction effects are again measured as deviations and we have:

$$\sum_{j=1}^{c} u_{12(ij)} = 0, \qquad \sum_{i=1}^{r} u_{12(ij)} = 0$$

that is

$$u_{12(i\cdot)} = 0, \qquad u_{12(\cdot j)} = 0$$

Estimation of the interaction effects would now be useful in identifying those categories responsible for any departure from independence. In terms of the interaction parameters the hypothesis of independence specifies that $u_{12(ij)} = 0$ for all values of i and j. Testing for independence is therefore seen to be equivalent to testing whether all the interaction terms in (5.11) are zero, or, in other words, that the model specified in (5.7) provides an adequate fit to the data. It should be noted that the model in (5.11) will fit the data perfectly since the expected values under this model are simply the observed frequencies. This arises because the number of parameters in the model is equal to the number of cell frequencies, and for this reason (5.11) is known as the *saturated model* for a two-dimensional table. The use of saturated models when analysing complex

multidimensional contingency tables is considered in Section 5.7.
Let us now examine the log-linear models available for a three-dimensional table. The saturated model in this case is:

$$\log_e F_{ijk} = u + u_{1(i)} + u_{2(j)} + u_{3(k)} + u_{12(ij)} + u_{13(ik)}$$
$$+ u_{23(jk)} + u_{123(ijk)} \qquad (5.12)$$

We see that this model includes main effect parameters for each variable, first-order interaction effect parameters for each pair of variables, and also parameters representing possible second-order effects between the three variables. The models corresponding to the various hypotheses connected with such a table discussed previously in Chapter 4 are obtained by equating certain of the terms in (5.12) to zero, but before considering this mention must be made that in this text we shall be restricting our attention to what Bishop terms *hierarchical models*. These are such that, whenever a higher order effect is included in the model, the lower order effects composed from variables in the higher effect are also included. Thus, for example, if a term u_{123} is included in a model, terms $u_{12}, u_{13}, u_{23}, u_1, u_2$, and u_3 must also be included. Consequently we shall regard a model such as:

$$\log_e F_{ijk} = u + u_{1(i)} + u_{2(j)} + u_{3(k)} + u_{123(ijk)} \qquad (5.13)$$

as not permissible.

The restriction to hierarchical models arises from the constraints imposed by the maximum likelihood estimation procedures, details of which are outside the scope of this text. In practice the restriction is of little consequence since most tables can be described by a series of hierarchical models, although in some cases this may require the table to be partitioned in some way (see Section 5.7).

Returning now to the model specified in (5.12), let us see how the hypothesis of no second-order interaction, previously expressed in terms of the probabilities p_{ijk} [see equation (4.14)], may be expressed in terms of the u-parameters. It is easy to show that the equivalent of (4.14) is:

$$H_0: u_{123(ijk)} = 0 \text{ for all } i, j, \text{ and } k$$

or, more simply (taking the alphabetic subscripts as understood):

$$H_0: u_{123} = 0 \qquad (5.14)$$

Other hypotheses may also be expressed in terms of the parameters in (5.12). For example, the hypothesis of mutual independence [see equation (4.4)], which specifies that there are no associations of any kind between the three variables, or, in other words, that there are no first-order interactions between any pair of variables, and no conjoint three-variable interaction, may now be written as follows:

$$H_0 : u_{12} = 0, \ u_{13} = 0, \ u_{23} = 0, \ u_{123} = 0 \qquad (5.15)$$

In this case our model for log-frequencies is simply:

$$\log_e F_{ijk} = u + u_{1(i)} + u_{2(j)} + u_{3(k)} \qquad (5.16)$$

which involves only an overall mean effect and main effect parameters for each of the three variables. If such a model provided an adequate fit to the data it would imply that differences between cell frequencies simply reflected differences between single variable marginal totals.

Now consider a model which specifies that $u_{12} = 0$ and therefore necessarily for hierarchical models also specifies that $u_{123} = 0$. The log-linear model would now be of the form:

$$\log_e F_{ijk} = u + u_{1(i)} + u_{2(j)} + u_{3(k)} + u_{13(ik)} + u_{23(jk)} \qquad (5.17)$$

Setting $u_{123} = 0$ is, as we have mentioned previously, equivalent to postulating that the interaction between variables 1 and 2 is the same at all levels of variable 3; setting $u_{12} = 0$ is equivalent to postulating that this interaction is zero. Consequently the model in (5.17) is seen to be specifying that there is no interaction between variables 1 and 2 at each level of variable 3 or, in other words, that variables 1 and 2 are conditionally independent given variable 3. In such a model variables 1 and 2 are each assumed to be associated with variable 3 since we have not specified that either u_{13} or u_{23} is zero.

Similarly the hypotheses of partial independence discussed in the preceding chapter can be considered in terms of (5.12) with some parameters set equal to zero. In this case we would specify u_{123} and one pair of u_{12}, u_{13}, and u_{23} to be zero. For example, hypothesis $H_0^{(1)}$ (see Section 4.5) is equivalent to the following log-linear model:

$$\log_e F_{ijk} = u + u_{1(i)} + u_{2(j)} + u_{3(k)} + u_{23(jk)} \qquad (5.18)$$

If we were to continue to delete u-terms from (5.12) so that there were fewer terms than in the complete independence model, that is (5.16), we would arrive at a model which did not include all variables. These are termed *non-comprehensive* models by Bishop, Feinberg, and Holland; if such a model were tenable for a set of data it would simply indicate that one (or more) of the variables were redundant and that the dimensionality of the table could be reduced accordingly. In practice, of course, only *comprehensive* models, that is those containing at least a main effect parameter for each variable, would be of concern.

Bishop, Feinberg, and Holland prove that a three-dimensional contingency table may be collapsed over any variable that is independent of at least one of the remaining pair, and the reduced table examined *without* the danger of misleading conclusions previously alluded to (see Section 4.3). Such a result shows that acceptance of one of the hypotheses of partial independence, that is showing that a model such as (5.18) provides an adequate fit to the data, would allow the table to be collapsed over *any* of the three variables with, consequently, a simplification of subsequent analyses. This procedure has been illustrated in the preceding chapter (see Section 4.5). The result also shows that, in the case where only conditional independence holds [that is, the model specified in (5.17)], care must be taken in deciding which variables are collapsible. For example, the spurious result found in collapsing Table 4.3 over the clinic variable is now explained, since *each* of the remaining two variables, amount of prenatal care and survival, is associated with this variable.

5.3. Fitting log-linear models and estimating parameters

In the preceding section we have seen that fitting particular log-linear models to the frequencies in a contingency table is equivalent to testing particular hypotheses about the table. Consequently, assessing the adequacy of a suggested model for the data follows exactly the same lines as that used in hypothesis testing, namely obtaining estimates of the theoretical frequencies to be expected assuming the model is correct, that is the values E_{ijk}, and comparing these with the observed values by means of either the χ_L^2 or the χ^2 statistic. The estimated expected values are obtained as for the equivalent hypothesis, and, as mentioned in Chapter 4, may in some cases be calculated explicitly from the relevant marginals of the observed values, but in other cases must be obtained using the type of iterative procedure to be discussed in Section 5.6.

A major advantage obtained from fitting log-linear models is that

TABLE 5.1. Two-dimensional data.

		Variable 2			
		1	2	3	
	1	20	56	24	100
Variable 1	2	8	28	14	50
	3	2	16	2	20
		30	100	40	170

we may obtain estimates of the parameters in the model. This allows us to quantify the effects of various variables and of interactions between variables. Estimates of the parameters in the fitted model are obtained as functions of the logarithms of the E_{ijk}, and the form of such estimates is very similar to those used for the parameters in analysis of variance models, as we shall see in the following example. Examining first the two-dimensional table shown in Table 5.1, let us compute the main effect parameters in the model:

$$\log_e F_{ij} = u + u_{1(i)} + u_{2(j)} \tag{5.19}$$

postulated for these data.

We first find the expected values under the model in (5.19). Since this model is equivalent to the hypothesis that the two variables are independent, these will be calculated using formula (1.9), and are given in Table 5.2. Estimates of the main effect parameters are now found by simply substituting the values in Table 5.2 for the F_{ij} in formulae (5.9) and (5.10). For example:

$$\hat{u}_{1(1)} = (\tfrac{1}{3})(\log_e 17.65 + \log_e 58.82 + \log_e 23.53) - (\tfrac{1}{9})(\log_e 17.65$$
$$+ \ldots + \log_e 4.71)$$
$$= 0.77$$

TABLE 5.2. Expected values for data of Table 5.1.

		Variable 2			
		1	2	3	
	1	17.65	58.82	23.53	100
Variable 1	2	8.82	29.41	11.76	50
	3	3.53	11.76	4.71	20
		30	100	40	170

TABLE 5.3. Estimated main effects for data in Table 5.1.

		Variable 1		Variable 2
	1	$\hat{u}_{1(1)} =$	0.77	$\hat{u}_{2(1)} = -0.50$
Category	2	$\hat{u}_{1(2)} =$	0.07	$\hat{u}_{2(2)} =$ 0.71
	3	$\hat{u}_{1(3)} = -0.84$		$\hat{u}_{2(3)} = -0.21$

The estimated main effects are shown in Table 5.3.

We see first that the estimates for each variable sum to zero; consequently the last effect may always be found by subtraction of the preceding ones from zero. Secondly we see that the size of the effects simply reflects the size of the marginal totals; that is, of the parameters $\hat{u}_{1(i)}$, $\hat{u}_{1(1)}$ is the largest since the first category of variable 1 has the largest marginal total amongst those of this variable. Similarly, of the parameters $\hat{u}_{2(j)}$, $\hat{u}_{2(2)}$ is largest. If we let $z_{ij} = \log_e E_{ij}$ and adopt a 'bar' notation for means, that is

$$\bar{z}_{i\cdot} = \frac{1}{c} \sum_{j=1}^{c} \log_e E_{ij} \text{ etc.,}$$

the main effect estimates may be written in the form taken by parameter estimates in the analysis of variance, that is:

$$\hat{u}_{1(i)} = \bar{z}_{i\cdot} - \bar{z}_{\cdot\cdot} \tag{5.20}$$

$$\hat{u}_{2(j)} = \bar{z}_{\cdot j} - \bar{z}_{\cdot\cdot} \tag{5.21}$$

Let us now return to the three-dimensional data in Table 4.1. In Section 4.5 the hypothesis that school behaviour (variable 1) is independent of school adversity (variable 2) and home condition variable 3) jointly was considered for these data. This hypothesis corresponds to the model:

$$\log_e F_{ijk} = u + u_{1(i)} + u_{2(j)} + u_{3(k)} + u_{23(jk)} \tag{5.22}$$

The expected values under the model are shown in Table 4.5, and from these we may now obtain estimates of the parameters in the model. Adopting again the nomenclature introduced above, that is letting $z_{ijk} = \log_e E_{ijk}$ etc., the estimates are given by:

$$\hat{u} = \bar{z}_{\ldots} \tag{5.23}$$

$$\hat{u}_{1(i)} = \bar{z}_{i\cdot\cdot} - \bar{z}_{\ldots} \tag{5.24}$$

$$\hat{u}_{2(j)} = \bar{z}_{\cdot j\cdot} - \bar{z}_{\ldots} \tag{5.25}$$

TABLE 5.4. Estimated interaction effects between school adversity and home condition for the data of Table 4.1.

$\hat{u}_{23(11)} = 0.392,$	$\hat{u}_{23(12)} = -0.392$
$\hat{u}_{23(21)} = -0.408,$	$\hat{u}_{23(22)} = 0.408$
$\hat{u}_{23(31)} = 0.016,$	$\hat{u}_{23(32)} = -0.016$

$$\hat{u}_{3(k)} = \bar{z}_{..k} - \bar{z}_{...} \tag{5.26}$$

$$\hat{u}_{23(jk)} = \bar{z}_{.jk} - \bar{z}_{.j.} - \bar{z}_{..k} + \bar{z}_{...} \tag{5.27}$$

The interaction parameters are shown in Table 5.4. Note that, since $\hat{u}_{23(j\cdot)} = 0$ and $\hat{u}_{23(\cdot k)} = 0$, only two of the six interaction parameters have to be estimated, the remaining four being determined by these relationships. The positive value of $\hat{u}_{23(11)}$ indicates a positive association between the low adversity school category and the not-at-risk for deviant behaviour category. Consequently more children will be found in this category than would be expected if the variables were independent. Of course, whether the difference is significant or simply attributable to random fluctuations will depend on whether or not $\hat{u}_{23(11)}$ is significantly different from zero; the question of the significance of parameter estimates will be discussed in Section 5.6. The pattern of values in Table 5.4 reflects the differences, already seen in Chapter 4 (Table 4.6), between the observed frequencies and the expected values calculated under the hypothesis of the independence of the two variables.

A difficulty which may arise when fitting log-linear models to contingency table data is the occurrence of zero cell entries. Such entries may arise in two ways; the first is when it is impossible to observe values for certain combinations of variables, in which case they are known as *a priori* zeroes and are discussed in the following chapter. Secondly, they may arise owing to sampling variation when a relatively small sample is collected for a table having a large number of cells; in this case zero cell entries are known as *sampling* zeroes. An obvious way to eliminate these is to increase the sample size. However, when this is not possible, it may be necessary, in some cases, to increase all cell frequencies by addition of a small constant, generally 0.5 before proceeding with the analysis. [Feinberg (1969) discusses more formal methods for determining the size of the constant to be added to each cell frequency to remove sampling zeroes.] The issues surrounding how to deal with sampling zeroes in general are discussed in detail in Bishop, Feinberg and Holland.

The analysis of multidimensional tables by the fitting of log-linear models requires, in general, a computer program to take the burden of the large amount of computation that may be involved especially when expected values have to be obtained iteratively. Some programs available for this purpose are described in Appendix B. For the analyses presented in later numerical examples Goodman's ECTA program was used.

5.4. Fixed marginal totals

In the preceding chapter it was mentioned that, corresponding to particular hypotheses, particular sets of expected value marginal totals are constrained to be equal to the corresponding marginal totals of observed values. In terms of the parameters in log-linear models this means that the u-terms included in the model determine the marginal constraints imposed on the expected values. For example, in a three-variable table, fitting a model including only main effects, that is:

$$\log_e F_{ijk} = u + u_{1(i)} + u_{2(j)} + u_{3(k)} \qquad (5.28)$$

fixes the following:

$$E_{i..} = n_{i..}, \ E_{.j.} = n_{.j.}, \ E_{..k} = n_{..k}$$

but no two-variable marginals are so constrained. In the case of the partial independence model, namely:

$$\log_e F_{ijk} = u + u_{1(i)} + u_{2(j)} + u_{3(k)} + u_{23(jk)} \qquad (5.29)$$

the following equalities hold:

$$E_{i..} = n_{i..}, \ E_{.j.} = n_{.j.}, \ E_{..k} = n_{..k}, \ and \ E_{.jk} = n_{.jk}$$

Now, for some sets of data certain marginal totals of observed frequencies are fixed by the sampling design, so the corresponding u-term *must* be included in the model so that the corresponding marginals of expected values are similarly fixed. For example, Table 5.5 shows some data previously considered by Bartlett (1935) and others, which gives the results of an experiment designed to investigate the propagation of plum root stocks from root cuttings. In this experiment the marginal totals $n_{.jk}$ are fixed *a priori* by the investigator at 240, and consequently, in any analysis of the data,

TABLE 5.5 Data on propogation of plum root stocks.

| | | Time of planting (Variable 2) | | | |
		At once		In Spring	
Length of cutting (Variable 3)		Long	Short	Long	Short
Condition of plant after experiment (Variable 1)	Alive	156	107	84	31
	Dead	84	133	156	209
		240	240	240	240

these marginals must be maintained at this value. Therefore, when fitting log-linear models to these data, *only* models including the term u_{23} would be considered. [For more details see Bishop (*op. cit.*).]

5.5. Obtaining expected values iteratively

As mentioned previously, expected values corresponding to some models cannot be obtained directly from particularly marginal totals of observed values. (This arises because, in these cases, the maximum likelihood equations have no explicit solution.) Consequently the expected values must be obtained in some other way. Bartlett (*op. cit.*) was the first to describe a method for obtaining expected values for a model that did not allow these to be obtained directly. More recently several authors, for example Bock (1972) and Haberman (1974), have suggested Newton–Raphson methods in this area. However, in this section we shall consider only the method of iterative proportional fitting originally given by Deming and Stephan (1940), and described later by Bishop (1969). To illustrate this procedure we shall apply it to the data in Table 4.1, obtaining expected values under the model which specifies only that there is no second-order interaction, namely:

$$\log_e F_{ijk} = u + u_{1(i)} + u_{2(j)} + u_{3(k)} + u_{12(ij)} + u_{13(ik)} + u_{23(jk)} \quad (5.30)$$

For a three-dimensional table the model in (5.30) is the only one for which expected values are not directly obtainable. [We know, of course, from the results discussed in Chapter 4, that such a model is acceptable for these data; it is used here simply as an illustration of obtaining expected values iteratively.]

Examining the terms in (5.30) we see that the totals $E_{ij\cdot}$, $E_{i\cdot k}$, and $E_{\cdot jk}$ are constrained to be equal to the corresponding observed marginals. The iterative procedure begins by assuming a starting value, $E_{ijk}^{(0)}$, for each E_{ijk}, of unity and proceeds by adjusting these proportionally to satisfy the first marginal constraint, namely that $E_{ij\cdot} = n_{ij\cdot}$, by calculating:

$$E_{ijk}^{(1)} = \frac{E_{ijk}^{(0)} \times n_{ij\cdot}}{E_{ij\cdot}^{(0)}} \tag{5.31}$$

[Note that $E_{ij\cdot}^{(1)} = n_{ij\cdot}$.]

The revised expected values $E_{ijk}^{(1)}$ are now adjusted to satisfy the second marginal constraint, that is $E_{i\cdot k} = n_{i\cdot k}$, as follows:

$$E_{ijk}^{(2)} = \frac{E_{ijk}^{(1)} \times n_{i\cdot k}}{E_{i\cdot k}^{(1)}} \tag{5.32}$$

[Note that $E_{i\cdot k}^{(2)} = n_{i\cdot k}$.]

The cycle is completed when the values given by (5.32) are adjusted to satisfy $E_{\cdot jk} = n_{\cdot jk}$, using:

$$E_{ijk}^{(3)} = \frac{E_{ijk}^{(2)} \times n_{\cdot jk}}{E_{\cdot jk}^{(2)}} \tag{5.33}$$

[Note that $E_{\cdot jk}^{(3)} = n_{\cdot jk}$.]

A new cycle now begins by using the values obtained from (5.33) in equation (5.31). The process is continued until differences between succeeding expected values differ by less than some small amount, say 0.01. Using this procedure on the data of Table 4.1, and remembering that we begin with initial values of unity in each cell, we have the following sequence of computations.

Cycle 1

Step 1. Using formula (5.31) we first obtain:

$$E_{111}^{(1)} = \frac{1 \times (16 + 15 + 5)}{(1 + 1 + 1)} = 12.00$$

Similarly

$$E_{112}^{(1)} = 12.00 \text{ and } E_{113}^{(1)} = 12.00;$$

$$E_{121}^{(1)} = \frac{1 \times (7 + 34 + 3)}{(1 + 1 + 1)} = 14.67$$

Similarly

$$E^{(1)}_{122} = 14.67 \text{ and } E^{(1)}_{123} = 14.67;$$

$$E^{(1)}_{211} = \frac{1 \times (1 + 3 + 1)}{(1 + 1 + 1)} = 1.67$$

Similarly

$$E^{(1)}_{212} = 1.67 \text{ and } E^{(1)}_{213} = 1.67;$$

$$E^{(1)}_{221} = \frac{1 \times (1 + 8 + 3)}{(1 + 1 + 1)} = 4.00$$

Similarly

$$E^{(1)}_{222} = 4.00 \text{ and } E^{(1)}_{223} = 4.00.$$

Step 2. Using formula (5.32) we now obtain:

$$E^{(2)}_{111} = \frac{12.00(16 + 7)}{(12.00 + 14.67)} = 10.35$$

$$E^{(2)}_{112} = \frac{12.00(15 + 34)}{(12.00 + 14.67)} = 21.05$$

$$E^{(2)}_{113} = \frac{12.00(5 + 3)}{(12.00 + 14.67)} = 3.60$$

TABLE 5.6. Expected values under the hypothesis of no second-order interaction for the data of Table 4.1.

| | | | Adversity of school condition | | | |
	Low		Medium		High	
Risk Index	N.A.R.	A.R.	N.A.R.	A.R.	N.A.R.	A.R.
	1. 12.00	1. 14.67	1. 12.00	1. 14.67	1. 12.00	1. 14.67
	2. 10.35	2. 12.65	2. 21.05	2. 26.95	2. 3.60	2. 4.40
N.D.	3. 16.13	3. 7.20	3. 15.60	3. 32.61	3. 4.52	3. 3.66
	F. 15.89	F. 7.11	F. 15.57	F. 33.28	F. 4.39	F. 3.61
Classroom behaviour						
	1. 1.67	1. 4.00	1. 1.67	1. 4.00	1. 1.67	1. 4.00
	2. 0.56	2. 1.41	2. 3.24	2. 7.76	2. 1.18	2. 2.82
D.	3. 0.87	3. 0.80	3. 2.40	3. 9.39	3. 2.48	3. 2.34
	F. 1.11	F. 0.89	F. 2.28	F. 8.72	F. 1.61	F. 2.39

1. Expected value at end of step 1.
2. Expected value at end of step 2.
3. Expected value at end of step 3.
F. Expected value at convergence on cycle four.

and so on. We would now proceed to step 3 by using formula (5.33) and then begin a new cycle. Table 5.6 shows the expected values obtained at the end of steps 1, 2, and 3 of cycle 1, and those at the completion of the process which in this case occurred on cycle 4.

In general this algorithm would operate by proportionally fitting the marginal totals fixed by the model. When expected values can be obtained explicitly from various marginal totals this algorithm is obviously not necessary, although it is easy to show that, for such cases, both methods will produce the same results. Bishop, Fienberg, and Holland (Ch. 3) discuss some rules for detecting when direct estimates are available; these may be especially useful to investigators without access to one of the computer programs previously mentioned.

5.6. Numerical example

Before proceeding with other topics connected with log-linear models, we shall examine in some detail the data in Table 5.7, with a view to illustrating some of the points covered in previous sections. These data concern coronary heart disease and were analysed by Ku and Kullback (1974). A total of 1330 patients have been categorized with respect to the following three variables:

Variable	Level
1. Blood pressure	1. Less than 127 mm Hg
	2. 127—146
	3. 147—166
	4. 167 +
2. Serum cholesterol	1. Less than 200 mg/100 cc
	2. 200—219
	3. 220—259
	4. 260 +
3. Coronary heart disease	1. Yes
	2. No

Let us first examine whether a model containing only main effects provides an adequate fit to these data. Such a model is equivalent to the hypothesis that the three variables are mutually independent, so the expected values may be calculated using formula (4.7); these are then compared with the observed frequencies by either the likelihood ratio criterion, χ_L^2, or the more usual χ^2 statistic. This

TABLE 5.7. Coronary heart disease data

| Serum cholesterol | | Blood Pressure | | | | |
		1	2	3	4	Total
	1	2	3	3	4	12
Coronary heard	2	3	2	1	3	9
disease	3	8	11	6	6	31
	4	7	12	11	11	41
	Total	20	28	21	24	93
	1	117	121	47	22	307
No coronary	2	85	98	43	20	246
heart disease	3	119	209	68	43	439
	4	67	99	46	33	245
	Total	388	527	204	118	1237
	Overall total	408	555	225	142	1330

results in $\chi_L^2 = 78.96$ and $\chi^2 = 99.54$. (In the rest of this chapter we shall give only the values of χ_L^2 when testing the fit of models.) To find the degrees of freedom associated with a particular model, we use:

d.f. = Number of cells in table − Number of parameters
$\qquad\qquad$ in fitted model that require estimating $\qquad$ (5.34)

In the case of a main effects model [see (5.28)] we have the following:

Parameters in model	Number of such parameters that require estimating	
Overall mean effect, u	1	
Main effect of variable 1, $u_{1(i)}$	$(r-1)$	[since $u_{1(\cdot)} = 0$]
Main effect of variable 2, $u_{2(j)}$	$(c-1)$	[since $u_{2(\cdot)} = 0$]
Main effect of variable 3, $u_{3(k)}$	$(l-1)$	[since $u_{3(\cdot)} = 0$]
	$r+c+l-2$	

Consequently from (5.34) we have:

$$d.f. = rcl - r - c - l + 2$$

This formula has been derived in Chapter 4 (see Section 4.7). For the data of Table 5.7, $r = 4$, $c = 4$, and $l = 2$, so the main effects model in this case has 24 d.f. The value of χ_L^2 obtained, namely 78.96, is therefore highly significant, and a log-linear model including only main effect parameters does not provide an adequate fit for these data.

The next model considered was that given in (5.30), which specifies that there is no second-order interaction between the three variables. In this case expected values have to be obtained iteratively using the procedure described in the preceding section. When this is done we may obtain χ_L^2 in the usual way to give $\chi_L^2 = 4.77$. In this case we have:

Parameters in model	Number of such parameters that require estimating
Overall mean effect, u	1
Main effect of variable 1, $u_{1(i)}$	$(r-1)$
Main effect of variable 2, $u_{2(j)}$	$(c-1)$
Main effect of variable 3, $u_{3(k)}$	$(l-1)$
Interaction effect, variables 1 and 2, $u_{12(ij)}$	$(r-1)(c-1)$ $\left[\text{since } u_{12(i\cdot)} = u_{12(\cdot j)} = 0\right]$
Interaction effect, variables 1 and 3, $u_{13(ik)}$	$(r-1)(l-1)$ $\left[\text{since } u_{13(i\cdot)} = u_{13(\cdot k)} = 0\right]$
Interaction effect, variables 2 and 3, $u_{23(jk)}$	$(c-1)(l-1)$ $\left[\text{since } u_{23(j\cdot)} = u_{23(\cdot k)} = 0\right]$
	$rc + rl + cl - r - c - l + 1$

and therefore the degrees of freedom of this model in this case is given by

$$d.f. = 32 - 4\cdot4 - 4\cdot2 - 4\cdot2 + 4 + 4 + 2 - 1$$
$$= 9$$

Consequently χ_L^2 is non-significant and there is no need to postulate any second-order interaction for these data. Table 5.8 shows the first-order interaction parameter estimates obtained for this model. In most cases when fitting models we shall be interested in the *simplest*, that is the one with fewest parameters that provides an adequate fit to the data. From the results above we can see that, for these data, we shall need a model somewhere between one involving only main effects and one involving all first-order interaction terms. An indication of which model might be appropriate may be obtained by examining the *standardized values* in Table 5.8. These values are obtained by dividing a parameter estimate by its standard error. Details of how to compute the later are given by Goodman (1971), who also shows that the standardized values have,

TABLE 5.8. Some parameter estimates for the 'no second-order interaction' model fitted to the data of Table 5.7. [Effects such as $\hat{u}_{13(12)}$ are obtained by using the fact that $\hat{u}_{13(1\cdot)} = 0$]

Variables		Magnitude of effect	Standard error	Standardized Value
1 and 3	$\hat{u}_{13(11)}$	− 0.219	0.111	− 1.967*
	$\hat{u}_{13(21)}$	− 0.238	0.106	− 2.248*
	$\hat{u}_{13(31)}$	0.075	0.117	0.637
	$\hat{u}_{13(41)}$	0.383	0.120	3.186*
2 and 3	$\hat{u}_{23(11)}$	− 0.227	0.125	− 1.810
	$\hat{u}_{23(21)}$	− 0.272	0.138	− 1.970*
	$\hat{u}_{23(31)}$	0.054	0.095	0.576
	$\hat{u}_{23(41)}$	0.445	0.090	4.949*
1 and 2	$\hat{u}_{12(11)}$	0.222	0.205	1.083
	$\hat{u}_{12(12)}$	0.111	0.233	0.474
	$\hat{u}_{12(13)}$	− 0.114	0.165	− 0.687
	$\hat{u}_{12(14)}$	− 0.219	0.159	− 1.381
1 and 2	$\hat{u}_{12(21)}$	− 0.018	0.202	− 0.091
	$\hat{u}_{12(22)}$	− 0.044	0.225	− 0.193
	$\hat{u}_{12(23)}$	0.155	0.148	1.045
	$\hat{u}_{12(24)}$	− 0.093	0.146	− 0.638
1 and 2	$\hat{u}_{12(31)}$	− 0.037	0.225	− 0.163
	$\hat{u}_{12(32)}$	0.027	0.245	0.112
	$\hat{u}_{12(33)}$	− 0.062	0.170	− 0.364
	$\hat{u}_{12(34)}$	0.071	0.159	0.448
1 and 2	$\hat{u}_{12(41)}$	− 0.167	0.234	− 0.716
	$\hat{u}_{12(42)}$	− 0.094	0.254	− 0.372
	$\hat{u}_{12(43)}$	0.020	0.170	0.120
	$\hat{u}_{12(44)}$	0.241	0.159	1.518

*Indicates a 'significant' effect

asymptotically, a standard normal distribution, and may therefore be compared with the normal deviate for any particular probability level, to obtain some idea as to the 'significance' of a particular effect; for example, they might be compared with the 5% normal deviate, namely ± 1.96.

Examination of Table 5.8 suggests that a model including only interaction terms u_{13} and u_{23} might provide an adequate fit for these data; consequently such a model, that is:

$$\log_e F_{ijk} = u + u_{1(i)} + u_{2(j)} + u_{3(k)} + u_{13(ik)} + u_{23(jk)} \qquad (5.35)$$

was tried. We see that it is the conditional independence model previously discussed and it implies, for these data, that there is no interaction between blood pressure and serum cholesterol level for both 'coronary heart disease' and 'no coronary heart disease' patients.

The expected frequencies for this model may be obtained explicitly from:

$$E_{ijk} = \frac{n_{i \cdot k} n_{\cdot jk}}{n_{\cdot \cdot k}} \qquad (5.36)$$

[see Lewis (*op. cit.*)]. The value of χ_L^2 was 24.40 with 18 d.f. compared with a value of chi-square from tables at the 5% level of 28.87. Therefore we may conclude that the model in (5.35) provides an adequate fit for the data of Table 5.7. Some of the estimated parameter values for this model are shown in Table 5.9. In general terms

TABLE 5.9. Some parameter estimates for the model specified in (5.35) fitted to the data of Table 5.7.

Variables		Magnitude	Standard error	Standardized value
1 and 3	$\hat{u}_{13(11)}$	-0.262	0.117	-2.233
	$\hat{u}_{13(21)}$	-0.247	0.105	-2.346
	$\hat{u}_{13(31)}$	0.084	0.117	0.720
	$\hat{u}_{13(41)}$	0.424	0.114	3.722
2 and 3	$\hat{u}_{23(11)}$	-0.247	0.124	-1.991
	$\hat{u}_{23(21)}$	-0.281	0.138	-2.029
	$\hat{u}_{23(31)}$	0.048	0.094	0.515
	$\hat{u}_{23(41)}$	0.480	0.090	5.333

the results indicate that there is a positive association between high blood pressure (that is level 4) and the presence of coronary heart disease [$\hat{u}_{13(41)} = 0.424$, $p < 0.05$], and similarly a positive association between high serum cholesterol level (that is category 4) and the presence of coronary heart disease [$\hat{u}_{23(41)} = 0.480$, $p < 0.05$]. In addition, the lack of a second-order interaction implies that (a) the interaction between blood pressure and coronary heart disease is the same at all levels of serum cholesterol, and (b) the interaction between serum cholesterol and heart disease is the same for all blood pressure levels.

5.7. Choosing a particular model

As the number of dimensions of a multidimensional table increases so does the number of possible models, and some procedures are obviously needed to indicate which models may prove reasonable to fit to the data and which are likely to be inadequate. One such procedure is to examine the standardized values in the saturated model for the data. These values may serve to indicate which parameters can be excluded and therefore which unsaturated models may be worth considering. In many cases, however, it will be found that several models provide an adequate fit to the data as indicated by the non-significance of the likelihood ratio criterion. In general the preferred model will be that with fewer parameters. In some cases, however, a test between rival models may be required to see which gives the most parsimonious summary of the data. Goodman (1971) and Fienberg (1970) show that for the hierarchical models considered in this chapter such a test may be obtained by subtracting the χ_L^2 values for the two models to assess the 'change' in goodness of fit which results from adding further parameters. For example, for the data in Table 5.7 we have seen that the model specified in (5.35) *and* that specified in (5.30) yield non-significant χ_L^2 values. The two models differ by the presence of the parameter u_{12} in the latter. The model including the extra parameter would only be preferred if it provided a significantly *improved* fit over the simpler model. The difference in the χ_L^2 values of the two models is 19.63. This may be compared with a chi-square with 9 d.f. (that is the difference in the degrees of freedom of the two models). Since this is significant we may conclude that addition of the parameter u_{12} to the model specified in (5.35) causes a significant improvement in fit, and consequently a model which includes first-order interactions between *all* pairs of variables is needed for these data.

It should be noted that in an *s*-dimensional table for which the *s*-factor effect is large, we cannot of course fit any unsaturated hierarchical model. In this case it may be informative to partition the table according to the levels of one of the variables and examine each of the resulting $(s-1)$-dimensional tables separately. Details are given in Bishop, Fienberg, and Holland, Ch. 4; these authors also describe other useful methods for choosing a particular log-linear model for a set of data.

5.8. Log-linear models for tables with ordered categories

The log-linear model is easily adapted to deal with contingency tables in which one or more of the variables have categories that fall into a natural order. The *u*-parameters used previously, which measured effects as deviations from an overall mean (and consequently summed to zero), are now replaced, for the ordered variables, by parameters representing linear, quadratic, and, if appropriate, higher order effects. The process is greatly simplified if the levels of the ordered variables can be assumed to be equally spaced, in which case *orthogonal polynomials* may be used. For example, for an ordered variable having three equally spaced categories, the linear and quadratic effects are obtained from the usual *u*-parameters by applying the orthogonal polynomial coefficients to be found, for example, in Kirk (1968), Table D.12, to give the following;

Linear main effect, $u_{1(L)}$.

$$u_{1(L)} = (\tfrac{1}{2})[(1)u_{1(1)} + (0)u_{1(2)} + (-1)u_{1(3)}]$$

$$= (\tfrac{1}{2})[u_{1(1)} - u_{1(3)}] \qquad (5.37)$$

assuming that the first level of the variable is the 'high' level.

Quadratic main effect, $u_{1(Q)}$.

$$u_{1(Q)} = (\tfrac{1}{4})[(1)u_{1(1)} - (2)u_{1(2)} + (1)u_{1(3)}]$$

$$= (\tfrac{1}{4})[u_{1(1)} + u_{1(3)} - 2u_{1(2)}] \qquad (5.38)$$

These effects will represent trends in the single variable marginal totals of the ordered variable. Similarly linear and quadratic effects may found for the interaction between the ordered and the other unordered variables; the size of such effects will indicate the similarity or otherwise of the trend of the ordered variable in

TABLE 5.10. Social class and number of years in present occupation at first attendance as out-patients for a sample of patients diagnosed as either neurotics or schizophrenics.

		Neurotic			Schizophrenic		
Duration of present occupation (Variable 1)		-1 yr	$2-5$ yr	$5+$ yr	-1 yr	$2-5$ yr	$5+$ yr
Social class (Variable 3)	I + II	4	6	18	5	10	8
	III	12	17	37	13	18	23
	IV + V	8	5	3	12	7	5
		24	28	58	30	35	36	211

Column header spanning: *Diagnosis (Variable 2)*

TABLE 5.11. Parameter estimates obtained from fitting a log-linear model specifying zero second-order interaction to the data in Table 5.10.

Variable		Magnitude of effect	Standard error	Standardized value
1	$u_{1(L)}$	0.164	0.104	1.582
	$u_{1(Q)}$	0.012	0.059	0.201
2	$u_{2(1)}$	-0.022	0.084	-0.262
3	$u_{3(1)}$	-0.215	0.123	-1.754
	$u_{3(2)}$	0.665	0.101	6.569
	$u_{3(3)}$	-0.450	0.130	-3.452
1 and 2	$u_{12(L1)}$	0.145	0.104	1.394
	$u_{12(Q1)}$	0.061	0.059	1.044
2 and 3	$u_{23(11)}$	0.071	0.123	0.578
	$u_{23(12)}$	0.075	0.101	0.739
	$u_{23(13)}$	-0.146	0.130	-1.117
1 and 3	$u_{13(L1)}$	0.350	0.153	2.286
	$u_{13(L2)}$	0.257	0.125	2.057
	$u_{13(L3)}$	-0.607	0.160	-3.791
	$u_{13(Q1)}$	-0.031	0.085	-0.370
	$u_{13(Q2)}$	0.017	0.071	0.241
	$u_{13(Q3)}$	0.014	0.092	0.155

different categories of the unordered variable. To clarify these ideas let us examine the data in Table 5.10. These data, collected in an investigation into the environmental causes of mental disorder, show the social class and number of years in present occupation for a sample of patients diagnosed either as neurotic or schizophrenic.

Variable 1 (duration of present occupation) falls into a natural order, and we wish to investigate the possible trends in this variable using log-linear models with linear and quadratic effect parameters. Fitting a model specifying zero second-order interaction results in the parameter estimates shown in Table 5.11 and a value of χ_L^2 of 2.00, which with 4 d.f. is non-significant. Examining first the main effects for variable 1, we see that only that representing linear trend approaches being significant; the quadratic effect is very small. These two effects reflect any trend in the overall number of patients in the three categories of the ordered variable, that is:

Duration of present occupation	− 1 yr	2–5 yr	5 + yr
Number of patients	54	63	94

The more interesting effects are those representing interactions between duration of present occupation and diagnosis or social class. Examining Table 5.11 we see that interactions between duration and diagnosis are not significant. This indicates that trends across duration are the same for both diagnoses. Interaction effects between duration and social class are however significant; the magnitude and sign of these effects show that there are *linear increases* in the number of people with increase in duration for social class categories 1 and 2, and a *linear decrease* for social class category 3. This is clearly seen in Table 5.12. The absence of any second-order relationship shows that the interaction effects between

TABLE 5.12. Number of people in the three categories of the duration variable, for the three levels of social class.

		Duration of present occupation		
		− 1 yr	2–5 yr	5 + yr
Social class	I + II	9	16	26
	III	25	35	60
	IV + V	20	12	8

social class and duration of present occupation are the same for both neurotics and schizophrenics.

5.9. Models for data for which one of the variables is A response variable

Several authors, for example Bhapkar and Koch (1968), have emphasized the difference between *factor* or *explanatory* variables which classify the unit of observation according to a description of the sub-population of the units to which he belongs (or to the experimental conditions which he undergoes), and *response* variables which classify according to a description of what happens to the unit during and/or after the experiment. For example, in the data of Table 5.5 we have a 'one response, two factor experiment', the response variable being the state of the plant after the experiment, that is alive or dead, and time of planting and length of cutting being factor variables. In any analysis of such data the primary interest will be in the 'effect' of the factor variables and combinations of these variables on the response variable.

To illustrate the models applicable in this case we shall use the data in Table 5.13, which results from a psychiatrist's interest in the extent to which patients' recovery is predictable from the symptoms they show when ill. For this purpose 819 male patients are assessed for the presence (+) or absence (−) of the three symptoms depression, anxiety, and delusions of guilt, and then at the end of a suitable time period each patient is rated as 'recovered' or 'not-recovered'.

TABLE 5.13. Incidence of symptoms amongst psychiatric patients.

		Depression (Variable 4)						
		−				+		
Anxiety (Variable 3)		−	+		−		+	
Delusions (Variable 2)		−	+	−	+	−	+	−	+
Condition of patient (Variable 1) — Recovered	68	3	58	3	70	23	129	59
Not recovered	137	3	70	3	69	10	87	27
Proportion recovered		0.3317	0.5000	0.4531	0.5000	0.5036	0.6970	0.5972	0.6860

For these data the first variable, namely condition of patient, is a response variable, and to investigate the effects of the explanatory variables on the recovery of patients one possibility would be to investigate differences in the rates of recovery for different symptoms and symptom combinations. (We shall assume during the rest of the discussion that response variables are, as in this example, dichotomous.) A simple method of making such an investigation would be to postulate models for the probability of recovery. For example, if we suspected that *none* of the symptoms had any effect on recovery, a possible model for the probabilities would be:

$$P_{jkl}^{234} = \theta \qquad (5.39)$$

where P_{jkl}^{234} is the probability of recovery for the cell arising from the jth category of variable 2 ($j = 1, 2$), the kth category of variable 3 ($k = 1, 2$), and the lth category of variable 4 ($l = 1, 2$). The observed proportions of patients recovering, p_{jkl}^{234}, given in Table 5.13 represent estimates of the probabilities. [in fact $E(p_{jkl}^{234}) = P_{jkl}^{234}$]. The model in (5.39) specifies that the probability of recovery is the same for all cells, that is for all combinations of symptoms. A more complicated model arises if we now introduce a parameter to represent the possible effect of the symptom *depression* on recovery, namely:

$$P_{jkl}^{234} = \theta + \theta_l \qquad (5.40)$$

Estimates of the parameters θ and $\theta_l (l = 1, 2)$ might now be obtained, and the model tested for goodness of fit. A problem arises, however, from the necessity that the probabilities satisfy the condition:

$$0 \leqslant P_{jkl}^{234} \leqslant 1 \qquad (5.41)$$

since parameter estimates could be obtained which lead to fitted probability values not satisfying (5.41). Because of this, and other problems (see Cox, 1970), models for probabilities are not generally considered. However, a convenient method does exist for representing the dependence of a probability on explanatory variables so that constraint (5.41) is inevitably satisfied, namely that of postulating models on the *logistic transformation* of the probability. This transformation is given by:

$$\lambda_{jkl}^{234} = \log_e \{ P_{jkl}^{234} / (1 - P_{jkl}^{234}) \} \qquad (5.42)$$

and since the probability varies from 0 to 1, the logistic varies from $-\infty$ to $+\infty$. Methods for fitting models to logistic functions are described by Dyke and Patterson (1952), Cox (1970), and

Maxwell and Everitt (1970). However, details of the methods given by these authors will not be considered here, since Bishop (1969) and Goodman (1971) show how the log-linear models for frequencies, discussed previously in this chapter, may be adapted to produce equivalent results. For example, let us suppose that we wish to postulate a model for the logistic function which involves only a single 'overall mean parameter', that is:

$$\lambda_{jkl}^{234} = \theta \tag{5.43}$$

Such a model is obviously equivalent to that specified in (5.39), although the parameters will not have the same values. In terms of the theoretical frequencies, F_{ijkl}, for Table 5.13:

$$P_{jkl}^{234} = F_{1jkl}/(F_{1jkl} + F_{2jkl})$$
$$= F_{1jkl}/F_{.jkl} \tag{5.44}$$

(The observed proportions of recovery are, of course, given by an equivalent relationship on the observed frequencies, n_{ijkl}, namely:

$$p_{jkl}^{234} = n_{1jkl}/n_{.jkl})$$

Substituting (5.44) in (5.42) we obtain:

$$\lambda_{jkl}^{234} = \log_e(F_{1jkl}/F_{2jkl}) \tag{5.45}$$

Suppose we now postulate a log-linear model for the frequencies, involving only an overall mean effect, and a main effect parameter for variable 1, namely:

$$\log_e F_{ijkl} = u + u_{1(i)} \tag{5.46}$$

If we now substitute (5.46) in (5.45) we obtain:

$$\lambda_{jkl}^{234} = u_{1(1)} - u_{1(2)} \tag{5.47}$$

which, since $u_{1(1)} + u_{1(2)} = 0$, becomes:

$$\lambda_{jkl}^{234} = 2u_{1(1)} \tag{5.48}$$

and this is equivalent to (5.43) with $\theta = 2u_{1(1)}$. Thus, fitting the log-linear model specified in (5.46) is equivalent to fitting the model for the logistic function given in (5.43). Let us now examine a more complicated example, namely that where we suppose that the symptom depression has an effect on recovery. The model for the logistic would be:

$$\lambda_{jkl}^{234} = \theta + \theta_l \tag{5.49}$$

An equivalent log-linear model is:

$$\log_e F_{ijkl} = u + u_{1(i)} + u_{4(l)} + u_{14(il)} \tag{5.50}$$

since substituting this in (5.45) gives:

$$\lambda_{jkl}^{234} = [u_{1(1)} - u_{1(2)}] + [u_{14(1l)} - u_{14(2l)}] \tag{5.51}$$

which is seen to be of the same form as (5.49) with $\theta = [u_{1(1)} - u_{1(2)}]$ and $\theta_l = [u_{14(1l)} - u_{14(2l)}]$. The parameter u_{14} should now be thought of as the 'effect' of the symptom depression on recovery. In a similar way other log-linear models can be shown to be equivalent to particular models for the logistic function. However, a complication arises when we consider parameter estimates. To illustrate this let us add a further parameter to the model in (5.50) namely any parameter *not* involving variable 1, say u_2, to give:

$$\log_e F_{ijkl} = u + u_{1(i)} + u_{2(j)} + u_{4(l)} + u_{14(il)} \tag{5.52}$$

Substituting this in (5.45) again gives (5.51), so (5.52) is also equivalent to the logistic model specified in (5.49). Inclusion of any other u-term, not involving variable 1, in the model for log-frequencies will also lead to (5.51); however, the inclusion of such terms will lead to different estimates of u_1 and u_{14}, and consequently of θ and θ_l; what is desirable is to choose that log-linear model that is equivalent to the particular logistic model under consideration *and* gives identical parameter estimates to those that would be obtained from fitting such models directly. Bishop (*op. cit.*) shows that the required log-linear model is that which includes the term $u_{IJKL\cdots(ijkl\cdots)}$ where I, J, K, L, etc. represent factor variables. Therefore the log-linear model needed to produce the same result as fitting the logistic model of (5.49) is:

$$\log_e F_{ijkl} = u + u_{1(i)} + u_{14(il)} + u_{234(jkl)} + \text{all other implied}$$
$$\text{lower order terms} \tag{5.53}$$

In this way the fitting of logistic models, as described by the authors previously mentioned, is included in the fitting of log-linear models.

5.9.1. Numerical example

Let us now examine the results of using these methods on the data of Table 5.13. First we shall fit the logistic model specified in (5.43) to these data; such a model implies that the incidence of recovery is the same for all symptom combinations. The required log-linear model is:

$$\log_e F_{ijkl} = u + u_{1(i)} + u_{234(jkl)} + \text{all other implied lower}$$
$$\text{order terms} \qquad (5.54)$$

Fitting this model by the methods outlined previously leads to a value of χ_L^2 of 50.44 with 7 d.f.; this is significant beyond the 1% level, and consequently we are led to its rejection.

Suppose now that we wish to test whether only the symptom *depression* has any effect on recovery. The model for the logistic function is that given in (5.49) and the required equivalent log-linear model is that in (5.53), which gives $\chi_L^2 = 14.87$ with 6 d.f. This is significant at the 5% level but not at the 1% level. For some purposes this would be an adequate fit; the estimate of the parameter $u_{14(11)}$ is -0.214 with standard error 0.081. This implies that incidence of recovery is significantly less amongst those patients *without* depression (that is category 1 of variable 4). In other words, prognosis is good for patients who are rated as having the symptom depression. If an improved fit was required we could go on to fit other effects in exactly similar fashion. Terms such as u_{123}, if needed in the model to provide an adequate fit, would now be indicative that variables 2 and 3 were not acting independently on the response variable, that is variable 1.

5.10. Summary

In this chapter we have discussed at some length the analysis of contingency tables by the fitting of log-linear models. Nevertheless we have covered only part of the area and the chapter should be regarded as an introduction to more complete work such as that of Bishop, Fienberg, and Holland. The techniques described allow a systematic approach to be taken to discovering the relationships that may be present in complex multidimensional tables, and provide a powerful addition to methods available for the analysis of contingency table data. In particular the estimates of first, second, and higher order interaction effects given by these methods are in general extremely useful, as we have seen, for assessing in advance how well a particular unsaturated model will fit the data and for indicating the relative importance of different interaction effects. They are also useful for detecting when the number of categories for a variable may be reduced; for details see Bishop, Fienberg, and Holland, Ch. 4.

CHAPTER SIX

Some special types of contingency table

6.1. Introduction

There are certain types of contingency table met with in practice
that require special consideration. In this chapter we shall discuss
some of the methods of analysis that have been suggested for such
tables, beginning with a description of the analysis of those tables
for which some cells have *a priori* zero entries.

6.2. Tables with *a priori* zeros

In the preceding chapter the problem of contingency tables having
empty cells was mentioned. It was indicated that, at least in principle,
'sampling zeros' could be made to disappear by simply increasing
the sample size; failing this they could easily be dealt with by the
addition of a small positive constant to each cell frequency. In many
situations, however, tables arise in which cells are empty *a priori*
owing to certain combinations of the variables being impossible,
that is having zero probability. Many authors call this type of
empty cell a 'structural zero'. Tables with structural zeros are
generally known as *incomplete* contingency tables, and their analysis
presents special problems, which have been considered by several
authors, including Goodman (1968), Mantel (1970), Fienberg (1972),
and Bishop, Fienberg, and Holland (1975), Ch. 5. Since much of this
work is outside the scope of the present text, we shall merely
illustrate, by means of an example, how such tables may be handled
by relatively straightforward extensions of the methods described
in the preceding chapter. For this purpose we shall use the data
in Table 6.1 given originally by Brunswick (1971) and also considered
by Grizzle and Williams (1972). These data arose from an investigation into health problems causing concern amongst teenagers.

TABLE 6.1. Data on teenagers' concern with health problems.

		Sex (Variable 2) Males		Females	
Age (Variable 1)		12–15	16–17	12–15	16–17
	S.	4	2	9	7
Health problem	M	—	—	4	8
causing concern	H.	42	7	19	10
(Variable 3)	N.	57	20	71	31

S : Sex reproduction. H : How healthy I am.
M : Menstrual problems. N : Nothing.

Since males were naturally not affected by menstrual problems, certain cells are *a priori* zero.

The analysis of such a table may proceed by fitting log-linear models from which parameters referring to cells containing structural zeros are excluded since they are known *a priori* to be zero. Expected values for such models must be obtained by using a simple modification of the algorithm described in Section 5.5; starting values for this algorithm are now taken as unity for non-empty cells, and zero for *a priori* empty cells; previously a starting value of unity was used for each cell. Use of these starting values ensures that the table of expected values found under a particular model will have zeros in the cells required. The observed and expected frequencies may now be compared by means of the familiar χ_L^2 statistic. The calculation of the correct number for the degrees of freedom is, however, complicated by the presence of the *a priori* empty cells. Previously, degrees of freedom were found from the equation [see equation (5.34)]:

$$\text{d.f.} = N_1 - N_2 \qquad (6.1)$$

where N_1 = number of cells in the table, and N_2 = number of parameters in the model that required estimating. In the case of incomplete tables the formula for degrees of freedom becomes:

$$\text{d.f.} = N_1 - N_2 - N_3 \qquad (6.2)$$

where N_3 = number of *a priori* empty cells. Care is needed here, however, in determining the number of parameters that need to be estimated, since those referring to the empty cells are known a

TABLE 6.2. Expected values under the complete independence model for the data shown in Table 6.1.

| | | Sex | | | |
		Males		Females	
Age		12–15	16–17	12–15	16–17
Health problem	S.	7.37	3.04	8.21	3.39
causing concern	M.	—	—	8.49	3.51
	H.	26.12	10.78	29.09	12.00
	N.	59.95	24.74	66.76	27.55

priori to be zero and must therefore be excluded. Returning to the data in Table 6.1, let us first fit the complete independence model; the expected values for this model are shown in Table 6.2, and the statistic χ_L^2 takes the value 28.24. The degrees of freedom in this case are:

$$16 - 2 - (1 + 1 + 1 + 3) = 8$$

since there are a total of two empty cells and the parameters that have to be estimated are u (grand mean effect), $u_{1(1)}$ (age effect), $u_{2(1)}$ (sex effect), and $u_{3(1)}, u_{3(2)}, u_{3(3)}$ (health concern effects). The obtained value of χ_L^2 is significant beyond the 1% level.

If we now fit a model specifying only that there is no second-order interaction, we obtain $\chi_L^2 = 2.03$. To determine the degrees of freedom let us consider the parameters in the model that need to be estimated, remembering that we are considering deviation parameters for which $u_{1(\cdot)} = 0, u_{2(\cdot)} = 0, u_{12(\cdot 1)} = 0$ etc.

Grand mean effect	u
Main effect, age	$u_{1(1)}$
Main effect, sex	$u_{2(1)}$
Main effect, health concern	$u_{3(1)}, u_{3(2)}, u_{3(3)}$
Interaction effect, age × sex	$u_{12(11)}$
Interaction effect, age × concern	$u_{13(11)}, u_{13(12)}, u_{13(13)}$
Interaction effect, sex × concern	$u_{23(11)}, \qquad , u_{23(13)}$

The interaction effect $u_{23(12)}$ is missing since it is *a priori* zero because of the empty cells. The number of parameters to be estimated is therefore 12, and the degrees of freedom is given by:

$$16 - 2 - 12 = 2$$

Consequently, our value of χ_L^2 is non-significant, and we may conclude that no second-order interaction need be postulated for these data.

6.3. Quasi-independence

The fitting of models with *a priori* zeros is closely allied to the problem of examining contingency tables for what Goodman (1968) has termed *quasi-independence* and which Fienberg (1972) has described as a form of independence *conditional* on the restriction of our attention to part of the table only. Testing tables for quasi-independence is often a useful way of identifying the sources of significance in an overall significant χ_L^2 or χ^2 value. To illustrate the concept we shall examine the data in Table 6.3 collected by Glass (1954) in a study of social mobility in Britain. (The expected values under the hypothesis that a subject's status is independent of his father's status are shown in parentheses alongside the observed values.)

The usual chi-square test of independence, that is of the hypothesis:

$$H_0 : p_{ij} = p_i.\, p_{.j}$$

gives $\chi^2 = 505.5$ which with 4 d.f. is highly significant; consequently the two classifications are not independent. In this example, however, there are various portions of the table that might be of interest for further investigation. For example, suppose that we were interested in investigating whether the data are compatible with the theory that, whilst there may be some 'status inheritance' from father to son in

TABLE 6.3. Cross-classification of a sample of British males according to each subject's status category and his father's status category.

| | | Subject's status | | |
		Upper	Middle	Lower
	Upper	588 (343.2)	395 (466.7)	159 (322.1)
Father's status	Middle	349 (453.8)	714 (617.0)	447 (439.1)
	Lower	114 (254.0)	320 (345.3)	411 (245.7)

(The expected values under the hypothesis that a subject's status is independent of his father's status are shown in parentheses alongside the observed values.)

TABLE 6.4. Data of Table 6.3 with diagonal elements excluded.

		Upper	Subject's status Middle	Lower
	Upper	—	395	159
Father's status	Middle	349	—	447
	Lower	114	320	—

every social stratum, once a son has moved out of his father's stratum his own status is independent of that of his father. This would entail testing for independence in that portion of the table given by excluding subjects having the same status as their fathers. Table 6.4, which is simply Table 6.3 with the diagonal entries excluded, shows the data that would now be of concern. Such a table may be thought of as arising from sampling from a form of truncated population from which sons having the same status as their father are excluded. The usual form of the hypothesis of independence given above must now be modified to reflect the fact that the diagonal elements of Table 6.4 are blank, by setting $p_{ii} = 0$ and adjusting the remaining probabilities so that they continue to sum to 1; our hypothesis now takes the following form:

$$H_0: \; p_{ij} = 0 \text{ if } i = j$$
$$= S p_{i.} p_{.j} \text{ if } i \neq j$$

where S is chosen so that:

$$\sum_{i=1}^{3} \sum_{j=1}^{3} p_{ij} = 1,$$

that is

$$S = \left(1 - \sum_{i=1}^{3} p_{i.} p_{.i} \right)^{-1}$$

This would now be termed a hypothesis of quasi-independence. In this case the maximum likelihood estimators of $p_{i.}$ and $p_{.j}$, and consequently of the expected frequencies, cannot be found directly. However, the latter can be found by using the iterative algorithm for finding expected values in a manner similar to that described in the preceding section, using starting values of unity for the cells in which we are interested, and zeros for those we wish to exclude. In this example we would therefore have:

$$E_{ij}^{(0)} = 1 \text{ if } i \neq j$$
$$= 0 \text{ if } i = j$$

where the $E_{ij}^{(0)}$'s are the starting values for the algorithm. Once the expected values under the hypothesis of quasi-independence have been found they may be compared with the observed frequencies by means of the usual χ^2 or χ_L^2 statistic. Using such a procedure on these data gives the expected values shown in Table 6.5. Comparing the observed off-diagonal cells with the expected values leads to $\chi^2 = 0.61$ which has a single degree of freedom (due to subtraction of 3 d.f. for the diagonal cells). This is not now significant and we may therefore conclude that the data are compatible with our previously expressed theory of social mobility.

The significance of the χ^2 value for the complete table is therefore due to discrepancies between the observed and expected values in those cells for which fathers and sons have the same class status. In each of these the observed frequency is greater than would be expected if the two classifications were independent. The greatest discrepancy is for the cell involving upper class sons with upper class fathers, and it might be of interest to test for independence when only this diagonal cell is excluded. Again we need to find expected values under this further hypothesis of quasi-independence using the fitting algorithm, with, in this case, the following starting values:

$$E_{ij}^{(0)} = 0 \text{ if } i = j = 1$$
$$= 1 \text{ otherwise}$$

Such a procedure results in a value of $\chi^2 = 143.42$. This has 3 d.f. and is highly significant; consequently 'status inheritance' does not occur only in the upper class stratum. If we now proceed one stage further and exclude also lower class sons having a lower class father, we find $\chi^2 = 20.20$ which with 2 d.f. is also significant.

TABLE 6.5. Estimated expected values under the hypothesis of quasi-independence, excluding subjects who have the same status as their fathers.

		Subject's status		
		Upper	Middle	Lower
	Upper	—	390.2	163.8
Father's status	Middle	353.8	—	442.2
	Lower	109.2	324.8	—

From these results it would appear that 'status inheritance' occurs in all three social strata.

6.4. Square contingency tables

Two-dimensional contingency tables in which the row and column variables have the same number of categories (say r) occur fairly frequently in practice and are known, in general, as *square tables*. They may arise in a number of different ways:

(I) When a sample of individuals is cross-classified according to two essentially similar categorical variables; for example, grade of vision of right and left eye.

(II) When samples of pairs of matched individuals, such as husbands and wives or fathers and sons, are each classified according to some categorical variable of interest.

(III) When two raters independently assign a sample of subjects to a set of categories.

For such tables hypotheses relating simply to independence are not of major importance. Instead we will be interested in testing for *symmetry* and *marginal homogeneity*. By symmetry in a square table we mean that:

$$p_{ij} = p_{ji} \ (i \neq j) \tag{6.3}$$

and by marginal homogeneity that:

$$p_{i.} = p_{.i} \ (\text{for } i = 1, 2, ..., r) \tag{6.4}$$

In a 2×2 table these are obviously equivalent; in larger tables symmetry as defined by (6.3) clearly implies marginal homogeneity as defined by (6.4). Chi-square tests for both symmetry and marginal homogeneity are available; we begin here with a description of the test for symmetry.

6.4.1. *Test for symmetry*

Examining again the data of Table 6.3 we see that it might be of interest to determine whether observations in cells situated symmetrically about the main diagonal have the same probability of occurrence, that is to test the symmetry hypothesis namely:

$$H_0: \ p_{ij} = p_{ji} \ (i \neq j)$$

In terms of these data such a hypothesis implies that changes in class between fathers and sons occur in both directions with the same probability. The hypothesis of symmetry has been considered by several authors, for example Grizzle et al. (1969), Maxwell (1970), and Bishop, Fienberg, and Holland, Ch. 8. Under this hypothesis we would expect the frequencies in the ijth and jith cells to be equal. The maximum likelihood estimate of the expected value in the ijth cell, E_{ij}, is given by:

$$
\begin{aligned}
E_{ij} &= \tfrac{1}{2}(n_{ij} + n_{ji}) \quad (i \neq j) \\
&= n_{ii} \quad\quad\quad\; (i = j)
\end{aligned}
\tag{6.5}
$$

Substituting these values in the usual form of the χ^2 statistic gives:

$$
\chi^2 = \sum_{i < j} (n_{ij} - n_{ji})^2 / (n_{ij} + n_{ji})
\tag{6.6}
$$

which under the hypothesis of symmetry has a chi-square distribution with $\tfrac{1}{2}r(r - 1)$ d.f. In the case of the data of Table 6.3 we have therefore:

$$
\begin{aligned}
\chi^2 &= \frac{(349 - 395)^2}{(349 + 395)} + \frac{(159 - 114)^2}{(159 + 114)} + \frac{(447 - 320)^2}{(447 + 320)} \\
&= 2.84 + 7.42 + 21.03 \\
&= 31.39
\end{aligned}
$$

This has three degrees of freedom and is highly significant; consequently the hypothesis of symmetry is rejected. The largest deviation occurs between frequencies n_{32} and n_{23}, and it appears that a larger number of sons of middle class fathers become lower class than the number of sons of lower class fathers who achieve middle class status; to a lesser extent the same is true for the number of sons of upper class fathers who become lower class, which is greater than the number of sons who go in the opposite direction.

6.4.2. Test for marginal homogeneity

In the case where the hypothesis of symmetry is rejected, the weaker hypothesis of marginal homogeneity may be of interest. This hypothesis postulates that the row marginal probabilities are equal to the corresponding column marginal probabilities. This may be formulated as follows:

$$
H_0: \; p_{i.} = p_{.i} \; (\text{for } i = 1, 2, ..., r)
$$

A test of this hypothesis has been given by Stuart (1955) and by

Maxwell (1970). For the general $r \times r$ table the test statistic is given by:

$$\chi^2 = \mathbf{d}'\mathbf{V}^{-1}\mathbf{d} \tag{6.7}$$

where $\mathbf{d}$ is a column vector of *any* $(r-1)$ of the differences $d_1, d_2, \ldots, d_r$, and $d_i = n_{i\cdot} - n_{\cdot i}$ and $\mathbf{V}$ is an $(r-1) \times (r-1)$ matrix of the corresponding variances and covariances of the d_i's and has elements:

$$v_{ii} = n_{i\cdot} + n_{\cdot i} - 2n_{ii} \tag{6.8}$$

$$v_{ij} = -(n_{ij} + n_{ji}) \tag{6.9}$$

(Stuart and Maxwell show that the same value of χ^2 results whichever value of d_i is omitted from the vector $\mathbf{d}$.)

If H_0 is true then χ^2 has a chi-square distribution with $(r-1)$ d.f. In the case of $r = 3$ Fleiss and Everitt (1971) show that the test statistic may be written as follows:

$$\chi^2 = \frac{\bar{n}_{23}d_1^2 + \bar{n}_{13}d_2^2 + \bar{n}_{12}d_3^2}{2(\bar{n}_{12}\bar{n}_{23} + \bar{n}_{12}\bar{n}_{13} + \bar{n}_{13}\bar{n}_{23})} \tag{6.10}$$

where

$$\bar{n}_{ij} = \tfrac{1}{2}(n_{ij} + n_{ji})$$

Applying formula (6.10) to the data of Table 6.3 for which $d_1 = -91$, $d_2 = -81$, $d_3 = 173$, and $\bar{n}_{12} = 372.0$, $\bar{n}_{13} = 136.5$, $\bar{n}_{23} = 383.5$ gives:

$$\chi^2 = 30.67$$

which with 2 d.f. is highly significant and we conclude that the marginal distribution of fathers' status differs from that of sons' status. The relevant marginal probabilities estimated from Table 6.3 are as follows:

	Class status		
	Upper	Middle	Lower
Fathers	0.33	0.43	0.24
Sons	0.30	0.41	0.29

The result indicates a general lowering in the class status of sons compared with that of their fathers.

Other tests of marginal homogeneity are given in Bishop, Fienberg, and Holland, Ch. 8; these authors also consider the concepts of symmetry and marginal homogeneity in the case of tables of more than two dimensions.

6.5. Summary

Methods for analysing special types of contingency table have been briefly described. Much fuller details are available in the papers cited. The reader should remember that techniques for dealing with incomplete tables and for testing for quasi-independence are subject to many technical problems not discussed here but which may cause difficulties in practice. The main purpose of this chapter has been to make readers aware of the existence of such methods, rather than to give a *detailed* account of their use.

Appendix A

Percentage points of the χ^2 distribution

| D.F. | *Probability (P)* | | | |
	0.050	0.025	0.010	0.001
1	3.841	5.024	6.635	10.828
2	5.991	7.378	9.210	13.816
3	7.815	9.348	11.345	16.266
4	9.488	11.143	13.277	18.467
5	11.071	12.833	15.086	20.515
6	12.592	14.449	16.812	22.458
7	14.067	16.013	18.475	24.322
8	15.507	17.535	20.090	26.125
9	16.919	19.023	21.666	27.877
10	18.307	20.483	23.209	29.588
11	19.675	21.920	24.725	31.264
12	21.026	23.337	26.217	32.909
13	22.362	24.736	27.688	34.528
14	23.685	26.119	29.141	36.123
15	24.996	27.488	30.578	37.697
16	26.296	28.845	32.000	39.252
17	27.587	30.191	33.409	40.790
18	28.869	31.526	34.805	42.312
19	30.144	32.852	36.191	43.820
20	31.410	34.170	37.566	45.315
21	32.671	35.479	38.932	46.797
22	33.924	36.781	40.289	48.268
23	35.173	38.076	41.638	49.728
24	36.415	39.364	42.980	51.179
25	37.653	40.647	44.314	52.620
26	38.885	41.923	45.642	54.052
27	40.113	43.194	46.963	55.476

D.F.		Probability (P)		
	0.050	0.025	0.010	0.001
28	41.337	44.461	48.278	56.892
29	42.557	45.722	49.588	58.302
30	43.773	46.979	50.892	59.703
40	55.759	59.342	63.691	73.402
50	67.505	71.420	76.154	86.661
60	79.082	83.298	88.379	99.607
80	101.879	106.629	112.329	124.839
100	124.342	129.561	135.807	149.449

Appendix B

The fitting of log-linear models to complex multidimensional contingency tables may involve a considerable degree of computation since expected values may need to be estimated using the iterative algorithm described in Chapter 5. Consequently computer programs are generally necessary for the analysis of such tables. Brief details of some such programs are given here.

(1) ECTA: Everymans Contingency Table Analysis.

This program was written by Professor Leo Goodman and is available by writting to him at: Department of Statistics, University of Chicago, 5734 University Avenue, Chicago, Illinois 60637.

The program fits log-linear models to contingency tables and gives estimates of effects and their standard errors. Both the χ^2 and χ_L^2 statistics are given as criteria for judging the fit of a model. Tables with *a priori* zeros and tests of quasi-independence can also be dealt with. The program is simple to use and is reasonably well documented.

(2) MULTIQUAL: log-linear analysis of nominal or ordinal qualitative data by the method of maximum likelihood.

This program was written by R. Darrell Bock and George Yates and is available from: National Educational Resources Inc., 215 Kenwood Avenue, Ann Arbor, Michigan 48103.

Again this program fits log-linear models to contingency tables and gives parameter estimates etc. It is, however, rather more powerful than ECTA and contains several more options. In many respects it is the qualitative data analogue of the program MULTI-VARIANCE produced by Jeremy Finn for the analysis of variance of quantitative data. It is more difficult to use than ECTA but is very well documented.

(3) GLIM: a FORTRAN program for fitting a class of generalized linear models.

This program was originally developed by J.A. Nelder and

details are available from: N.A.G., Central Office, Oxford University, Computer Laboratory, Banbury Road, Oxford.

This is a very comprehensive program for the analysis of contingency tables. It has a simple interpretive free-format language which allows data specification, data input, maximal model specification, and a directive which causes a declared model to be fitted. An interactive version of the program is available which is easy to use and enables the exploration of complex tables to be undertaken in an extremely flexible manner.

(4) CATLIN: a FORTRAN program for the analysis of contingency tables.

This program has been developed at the University of North Carolina and implements the methods for the analysis of categorical data described by Grizzle, Starmer, and Koch (1969). It can be an extremely useful program for the analysis of contingency tables of all types including those with missing cells etc. It does, however, suffer from the disadvantage of requiring the user to supply a design matrix. This makes the use of this program rather more difficult than the others mentioned.

Bibliography

Armitage, P. (1971), *Statistical Methods in Medical Research*, Blackwell Scientific Publications, Oxford and Edinburgh.

Barlow's Tables, (1957), L.J. Comrie (ed.), Eand F.N. Span Ltd. London.

Barlett, M.S. (1935), Contingency Table interactions, *J. Roy. Statist. Soc. Suppl.*, **2**, 248–252.

Bennett, B.M. and Hsu, P. (1960), On the power function of the exact test for the 2×2 contingency table, *Biometrika*, **47**, 393–397.

Berkson, J. (1946), Limitations of the application of the fourfold table analysis to hospital data, *Biometrics*, **2**, 47–53.

Bhapkar, V.P. (1968), On the analysis of contingency tables with a quantitative response, *Biometrics*, **24**, 329–338.

Bhapkar, V.P. and Koch, G.G. (1968), Hypothesis of 'no interaction' in multidimensional contingency tables, *Technometrics*, **10**, 107–123.

Birch, M.W. (1963), Maximum likelihood in three-way contingency tables, *J. Roy. Statist. Soc.* (series B), **25**, 220–223.

Birtchnell, J. and Alarcon, J. (1971), Depression and attempted suicide, *Brit. J. Psychiatry*, **118**, 289–296.

Bishop, Y.M.M. (1969), Full contingency tables, logits and split contingency tables, *Biometrics*, **25**, 383–399.

Bishop, Y.M.M., Fienberg, S.E. and Holland, F.W. (1975), *Discrete Multivariate Analysis*, Massachusetts Institute of Technology Press.

Bock, R.D. (1972), Estimating item parameters and latent ability when responses are scored in two or more nominal categories, *Psychometrika*, **37**, 29–51.

Brunden, M.N. (1972), The analysis of non-independent 2×2 tables using rank sums, *Biometrics*, **28**, 603–607.

Brunswick, A.F. (1971), Adolescent health, sex and fertility, *Amer. J. Public Health*, **61**, 711–720.

Chapman, D.G. and Jun-Mo Nam (1968), Asymptotic power of chi-square tests for linear trends in proportions, *Biometrics*, **24**, 315–327.

Cochran, W.G. (1954), Some methods for strengthening the common χ^2 tests, *Biometrics*, **10**, 417–451.

Conover, W. J. (1968), Uses and abuses of the continuity correction, *Biometrics*, **24**, 1028.

Conover, W.J. (1974), Some reasons for not using the Yates continuity correction on 2 × 2 contingency tables, *J. Amer. Statist. Assoc.*, **69**, 374–376.

Cox, D.R. (1970), *Analysis of Binary Data*, Methuen, London.

Cramer, H. (1946), *Mathematical Methods for Statistics*, Princeton Univ. Press.

Darroch, J.N. (1962), Interactions in multi-factor contingency tables, *J. Roy. Statist. Soc.* (series B), **24**, 251–263.

Deming, W.E. and Stephan, F.F. (1940), On a least squares adjustment of a samples frequency table when the expected marginal totals are known, *Ann. Math. Statist.*, **11**, 427–444.

Dyke, G.V. and Patterson, H.D. (1952), Analysis of factorial arrangements when the data are proportions, *Biometrics*, **8**, 1–12.

Feldman, S.E. and Klinger, E. (1963), Short-cut calculation of the Fisher-Yates "exact test", *Psychometrika*, **28**, 289–291.

Fienberg, S.E. (1969), Preliminary graphical analysis and quasi-independence for two-way contingency tables, *Applied Statistics*, **18**, 153–168.

Fienberg, S.E. (1970), The analysis of multidimensional contingency tables, *Ecology*, **51**, 419–433.

Fienberg, S.E. (1972), The analysis of incomplete multi-way contingency tables, *Biometrics*, **28**, 177–202.

Finney, D.J., Latscha, R., Bennet, B.M. and Hsu, P. (1963), *Tables for Testing Significance in a 2 × 2 Contingency Table*, Cambridge University Press, London.

Fisher, R.A. (1950), *Statistical Methods for Research Workers*, Oliver and Boyd, Edinburgh.

Fisher, R.A. and Yates, F. (1957), *Statistical Tables for Biological, Agricultural and Medical* (5th Ed.), Oliver and Boyd, Edinburgh.

Fleiss, J.L. (1973), *Statistical Methods for Rates and Proportions*, Wiley, New York.

Fleiss, J.L. and Everitt, B.S. (1971), Comparing the marginal totals of square contingency tables, *Brit. J. Math. Statist. Psychol.*, **24**, 117–123.

Gail, M. and Gart, J. (1973), The determination of sample sizes for use with the exact conditional test in 2 × 2 comparative trials, *Biometrics*, **29**, 441–448.

Gart, J.J. (1969), An exact test for comparing matched proportions in cross-over designs, *Biometrika*, **56**, 75–80.

Glass, D.V. (1954), *Social Mobility in Britain*, Free Press, Glencoe, Illinois.

Goodman, L.A. (1968), The analysis of cross-classified data: independence, quasi-independence, and interactions in contingency tables with or without missing data, *J. Amer. Statist. Assoc.*, **63**, 1091–1131.

Goodman, L.A. (1970), The multivariate analysis of qualitative data: interactions among multiple classification, *J. Amer. Statist. Assoc.*, **65**, 226–256.

Goodman, L.A. (1971), The analysis of multidimensional contingency tables:

Stepwise procedures and direct estimation methods for building models for multiple classification, *Technometrics*, **13**, 33–61.

Goodman, L.A. and Kruskal, W.H. (1954), Measures of association for cross-classifications, Part I, *J. Amer. Statist. Assoc.*, **49**, 732–764.

Goodman, L.A. and Kruskal, W.H. (1959), Measures of association for cross-classifications, Part II, *J. Amer. Statist. Assoc.*, **54**, 123–163.

Goodman, L.A. and Kruskal, W.H. (1963), Measures of association for cross-classifications, Part III, Approximate sampling theory, *J. Amer. Statist. Assoc.*, **58**, 310–364.

Goodman, L.A. and Kruskal, W.H. (1972), Measures of association for cross-classifications, Part IV, Simplification of asymptotic variances, *J. Amer. Statist. Assoc.*, **67**, 415–421.

Grizzle, J.E. and Williams, O. (1972), Log-linear models and tests of independence for contingency tables, *Biometrics*, **28**, 137–156.

Grizzle, J.E., Starmer, F. and Koch, G.G. (1969), Analysis of categorical data by linear models, *Biometrics*, **25**, 489–504.

Haberman, S.J. (1973), The analysis of residuals in cross-classified tables, *Biometrics*, **29**, 205–220.

Haberman, S.J. (1974), *The Analysis of Frequency Data*, Univ. of Chicago Press, Chicago.

Hays, W.L. (1973), *Statistics for the Social Sciences*, Holt, Reinhart and Winston, New York.

Irwin, J.O. (1949), A note on the subdivision of χ^2 into components, *Biometrika*, **36**, 130–134.

Kendall, M.G. (1952), *The Advanced Theory of Statistics*, Vol. 1, Griffin, London.

Kendall, M.G. (1955), *Rank Correlation Methods*, Hafner, New York.

Kendall, M.G. and Stuart, A. (1961), *The Advanced Theory of Statistics*, Griffin, London.

Kimball, A.W. (1954), Short-cut formulae for the exact partition of χ^2 in contingency tables, *Biometrics*, **10**, 452–458.

Kirk, R.E. (1968), *Experimental Design Procedures for the Behavioural Sciences*, Brooks/Cole Publishing Co., California.

Ku, H.H. and Kullback, S. (1974), Log-linear models in contingency table analysis, *Amer. Statistician*, **28**, 115–122.

Lancaster, H.O. (1949), The derivation and partition of χ^2 in certain discrete distributions, *Biometrika*, **36**, 117–129.

Lewis, B.N. (1962), On the analysis of interaction in multidimensional contingency tables, *J. Roy. Statist. Soc.* (series A), **125**, 88–117.

Lewontin, R.C. and Felsenstein, J. (1965), The robustness of homogeneity tests in $2 \times N$ tables, *Biometrics*, **21**, 19–33.

Lindley, D.V. and Miller, J.C.P. (1953), *Cambridge Elementary Statistical Tables*, Cambridge University Press, London.

Mantel, N. (1970), Incomplete contingency tables, *Biometrics*, **26**, 291–304.

Mantel, N. (1974), Comment and a suggestion on The Yates continuity correction, *J. Amer. Statist. Assoc.*, **69**, 378–380.

Mantel, N. and Greenhouse, S.W. (1968), What is the continuity correction?, *Amer. Statistician*, **22**, 27–30.

Mantel, N. and Haenszel, W. (1959), Statistical aspects of the analysis of data from retrospective studies of disease, *J. Natl. Cancer Inst.*, **22**, 719–748.

Maxwell, A.E. (1970), Comparing the classification of subjects by two independent judges, *Brit. J. Psychiatry*, **116**, 651–655.

Maxwell, A.E. (1974), Tests of association in terms of matrix algebra, *Brit. J. Math. Statist. Psychol.*, **26**, 155–166.

Maxwell, A.E. and Everitt, B.S. (1970), The analysis of categorical data using a transformation, *Brit. J. Math. Statist. Psychol.*, **23**, 177–187.

McNemar, Q. (1955), *Psychological Statistics*, Wiley, New York.

Mood, A.M. and Graybill, F.A. (1963), *Introduction to the Theory of Statistics*, McGraw-Hill, New York.

Pearson, K. (1904), *On the Theory of Contingency and Its Relation to Association and Normal Correlation*, Drapers' Co. Memoirs, Biometric Series No. 1, London.

Rodger, R.S. (1969), Linear hypothesis in $2 \times a$ frequency tables, *Brit. J. Math. Statist. Psychol.*, **22**, 29–48.

Roy, S.N. and Kastenbaum, M.A. (1956), On hypothesis of no interaction in a multiway contingency table, *Ann. Math. Statist.*, **27**, 749–751.

Slakter, M.J. (1966), Comparative validity of the chi-square and two modified chi-square goodness of fit tests for small but equal expected frequencies, *Biometrika*, **53**, 619–623.

Somers, R.H. (1962), A new asymmetric measure of association for ordinal variables, *Amer. Sociological Rev.*, **27**, 799–811.

Stuart, A. (1955), A test for homogeneity of the marginal distribution in a two-way classification, *Biometrika*, **42**, 412–416.

Taylor, I. and Knowelden, J. (1957), *Principles of Epidemiology*, Churchill, London.

Williams, E.J. (1952), Use of scores for the analysis of association in contingency tables, *Biometrika*, **39**, 274–289.

Williams, K. (1976), The failure of Pearson's goodness of Fit Statistic. *The Statistician*, **25**, 49.

Yates, F. (1934), Contingency tables involving small numbers and the chi-square test, *J. Roy. Statist. Soc. Suppl.*, **1**, 217–235.

Index